Acknowledgments

This curriculum was developed through funds from a USDA-NIFA Grant (Award # 2012-68003-19621) titled "Development of Novel Salmonella Control Practices and Integrated Education Program to Reduce Salmonellosis." Research **Principal Investigators** include Dr. Hosni Hassan, Dr. Matthew Koci at North Carolina State University, and Dr. Andrea Azcarate-Peril at the University of North Carolina at Chapel Hill. Extension and education components for this grant were provided by North Carolina 4-H and the North Carolina Kenan Fellows Program, both at North Carolina State University.

Also instrumental in the creation of this curriculum:

Kenan Fellow Authors
Lauren Boop
Brad Rhew
Cassandra Palmer

4-H Curriculum Specialist
Amy Chilcote

4-H Evaluation Specialist
Autumn Guin

4-H Microbiology Program Coordinator
Anne Greenaway

4-H Graphic Design/Illustration
Tabitha Barnes

4-H Illustrator
Uyen Nguyen- NCSU Chemistry Undergraduate

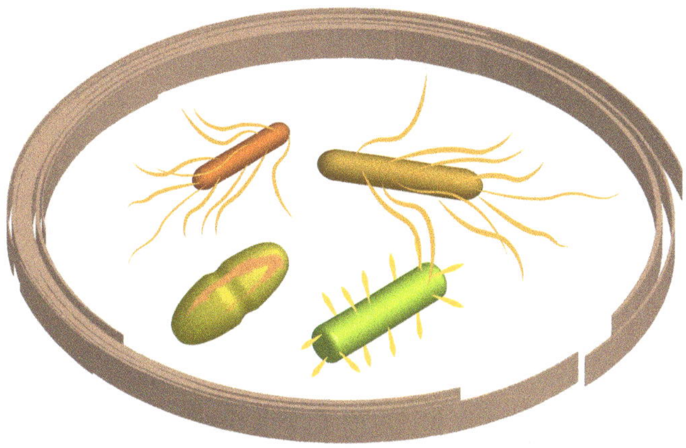

4-H curriculum is vetted through both formal and non formal educational environments. This curriculum has been piloted through the NC 4-H Curriculum Guidelines and evaluated through the following youth directed systems: summer camp, school enrichment, and 4-H community clubs. The following counties assisted in piloting this curriculum: Bladen, Cabarrus, Cherokee, Currituck, Iredell, Johnston, Lincoln, Nash, Rowan, Sampson, Stanley, and Warren County Cooperative Extensions.

Your purchase helps ensure that this curriculum will stay current with state and national educational standards as well as provide additional curricula materials locally to North Carolina 4-H agents, youth, and volunteers.

Introduction "Microbes on My Mind- A Middle School Perspective"

As you are reading this you have already come in contact with millions of microbes. Even though you cannot see them, they are all around you. This unit gives middle school participants an insight into just what those tiny microbes are and where they can be found.

In this unit, participants will discover the role microbes play in our bodies, our ecosystem, and our food. The activities provided will make it easy for the facilitator to explain the impact microorganisms have on humans in an interactive and engaging experience. These lessons are designed to encourage curiosity, resilience, and cooperation in our learners that will allow them to develop an understanding of the relationship between food safety, microbiology, and diseases, as well as the careers associated with microbiology. Participants will make connections between their existing knowledge of microorganisms, human health, and potential career paths related to this topic. Through the activities in this unit, participants will deepen their understanding of microbes in history, where they can be found, how to prevent the spread of microbes on them and in their food, and how certain careers have advanced to provide a better understanding of microorganisms. Participants will walk away knowing that microbes are a part of their everyday life and that not all microbes are harmful, but some can also be beneficial.

Middle school participants will address some of the misconceptions surrounding microbiology and gain a better understanding of how important microbes are in our ecosystem. Participants will gain an understanding of the importance of food safety and the precautions that need to be taken to keep their bodies healthy and protected.

ADDITIONAL MATERIALS

There are additional materials for each lesson available on our NC 4-H curriculum website (nc4hcurriculum.org).

Content for lessons 1-6:

- **Lesson variations for remote learning**
- **Videos**
- **PowerPoint presentations**
- **Interactive content**
- **PDFs separated by activity for easy printing**
- **Fillable PDFs for online submission of worksheets**
- **New content added regularly**

Visit NC4HCurriculum.org under the Virtual Resources menu section.

***Bonus access* to all "Grab and Go" content on the menu under Grab and Go. Those are not lessons from this microbiology curriculum.**

Access Code for "*Microbes On My Mind*" and all "*Grab and Go*" content:

60ad40c9de0bc

Table of Contents

Middle School Microbiology Unit
Scope and Sequence Variations

Teaching Lessons over 5 days for 1 ½-2 hours a day:

DAY	LESSONS
1	Lesson 1: The Microbes in You: An Introduction to Microbiology Lesson 5: Microbes All Around Us—Start the Plating activity
2	Lesson 3: Microbes in our Food (Complete the yogurt making part) Lesson 5: Observe bacteria cultures
3	Lesson 3: Observe Yogurt and complete lesson Lesson 4: Microbes in Our Water Lesson 5: Observe bacteria cultures
4	Lesson 2: Microbes in History Lesson 5: Observe bacteria cultures
5	Lesson 6: DNA Extraction Activity Lesson 5: Complete last observation of bacteria cultures and the rest of the lesson

Teaching Lessons over 2 days: 3-5 hours a day:

DAY	LESSONS
1	Lesson 1: The Microbes in You: An Introduction to Microbiology Lesson 3: Microbes in our Food (Complete the yogurt making part) Lesson 4: Microbes in our Water Lesson 2: Microbes In History (start)
2	Lesson 3: Complete "Most Wanted" posters if time is needed Lesson 2: Observe yogurt and complete lesson Lesson 5: Microbes All Around Us: Have Petri Dishes already prepared for participants to observe since time is not permitted to let bacteria to grow Lesson 6: DNA Extraction Activity

Teaching all lessons in 3 days: 2-3 hours a day:

DAY	LESSONS
1	Lesson 1: The Microbes in You: An Introduction to Microbiology Lesson 5: Microbes All Around Us: Allow participants to create bacteria cultures but also have some prepared if there is not enough time for growth Lesson 4: Microbes in Our Water
2	Lesson 5: Observe bacteria cultures and do activities Lesson 3: Microbes in Our Food: (Complete the Yogurt Making Part) Lesson 2: Microbes in History
3	Lesson 2: Microbes in History –finish if needed Lesson 3: Observe yogurt and complete activities Lesson 6: DNA Extraction Activity Lesson 5: Observe bacteria cultures and complete all activities with lesson

Middle School Microbiology Unit
Scope and Sequence Variations Continued

Teaching all lessons in 1 day: 6-7 hours a day:

DAY	LESSON SEQUENCE
1	**If curriculum is only taught in one day, these are the top 4 suggested lessons to cover: Lesson 3: Microbes in our Food: Complete yogurt making—should be ready by end of day but have some prepared Lesson 1: The Microbes in You: An Introduction to Microbiology Lesson 4: Microbes in Water Lesson 6: DNA Extraction Activity Lesson 3: Check Yogurt to see if it has formed and complete the conclusion part of lesson

Experiential Learning Model

A Holistic Approach to Learning

Experiential Learning is youth-centered. It is an active learning process, which first involves youth in an activity, asks them to reflect on the activity to determine what was useful or important to remember, and then tasks youth with understanding how the information they learned applies to their lives now and in the future. This learning process is holistic and allows youth to absorb, apply, and build upon life experiences.

Pfeiffer and Jones' experiential learning model is commonly used in 4-H. However, other pioneers such as Kolb provide a wealth of research that emphasizes the importance of the adult facilitator. According to Kolb, the experiential learning model allows the learner actively to participate in what is learned. The facilitator's role is to connect with the youth to spark their interest and intrinsic motivation so that the youth can gain, revise, or enhance their life skills. The facilitator plays multiple roles during this process: balancing the attention of the learner, engaging with the subject matter, and creating a safe and holistic learning environment.

Tips to think about when creating a welcoming environment for learning:

- Patience is a virtue. When waiting for youth to respond fully, having patience and allowing time for them to think is important.
- You are not the judge. Accept all responses.
- Experiential questions are open-ended to allow for expression. Make sure that you acknowledge youth who respond to you. Not acknowledging all youth can stop the flow of answers and make youth less likely to respond in the future.
- You are not the young person. Don't answer the question.
- Encourage deep conversation. If a young person gives a short response, provide follow-up questions to allow for deeper conversation. Examples could be: "Would you provide an example of what you mean? Can you tell me more about your thoughts?"
- Avoid yes or no questions. These are not open-ended and stop the flow of conversation.
- Don't lead the youth to answers you want to hear.
- Allow the youth to share their responses in pairs rather than with the entire group.

1. Experience: Let's Do It!!!
   Key Reminders to Leader:
- "Sit on Your Hands"
- Observe
- Facilitate to the "bigger picture"

Key Objectives for Discovery:
- To Explore
- To Examine
- To Construct
- To Arrange

2. Share
Key Concept: Respond
Key Question: "What Happened?"

3. Reflect
Key Concept: Analyze Patterns
Key Question: "What's Important?"

Leader Role
- Allow adequate processing time for sharing.
- Use open-ended questions to stimulate thinking and feeling.
- Encourage "pair share" and large group share.

4. Generalize
Key Concept: Inference
Key Question: "So What?"

5. Apply
Key Concept: Application
Key Question: "Now What?"

Leader Role
- Facilitate youth finding ways to use what they have learned in new situations.
- Guide youth in making connections between the inner meaning of the activity and the broader world.

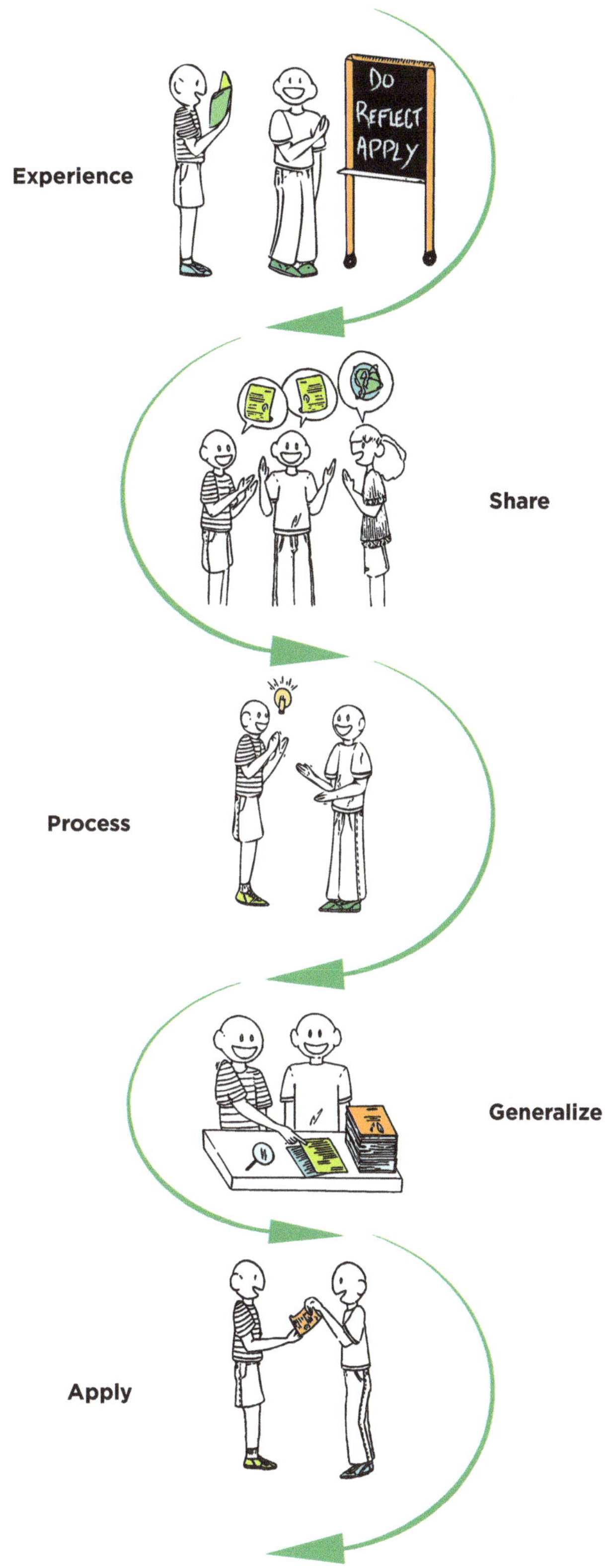

Share Questions (What Happened?)
What did you like about this activity?
What part of the activity was the hardest to do?
How did you decide what to choose?
What was the most fun about doing the activity?
How did your group work together?
How did you feel about successfully being able to____?
What are some decisions you had to make to carry out this activity?

What steps did you go through before you made your decision?
Why do you think everyone in your group didn't agree about everything?
What was it like to have to make quick decisions?
How did you keep track of everyone's ideas?
Do you think you get more ideas working alone or in a group? Why?
What kind of feelings did you have when group members argued?
What did you observe about the way the groups disagreed (or agreed)?
How did you use your various senses to ___?
What new words did you learn?
Why do you think people have different ideas about what is correct?

Reflect Questions (What's Important?)
What did you learn about yourself by doing this activity?
Why is it important to know___?
How did your group decide ___?
What did you learn as a group that you might not have learned alone?
What were some common themes or thoughts you heard?
How were each person's viewpoints the same? How were they different?
What problems came up over and over?
What would you do if_____ were to happen?
What was easy or difficult about working with a group to____?
What did you do if everyone in the group didn't agree on___?
How did this differ from the way you are usually taught in school?
Give an example of a challenge you had and what you did to solve it.
What works best to get people involved and excited about doing this type of activity?
Why is learning with others sometimes more fun than learning alone?
What suggestions would you have for someone who wanted to____?

Generalize Questions (So What?)
What other situations like this have you experienced?
Why is it important that each person has his/her own view?
When have you had to___ before?
Where can you find resources to help you make some of your decisions about__?
When else have you had fun and learned new things at the same time?
Why is it important to have plenty of information before making decisions?
When do you make decisions that require everyone in the group to agree?
What do you do when you don't agree with the group?
What did you learn about your own skill in making decisions?
What did you learn about your own skill in communicating with others?
Describe five ways in which new ideas can be communicated to you.
In what ways do people help each other learn new things?
What are some ways you like to learn?
What did you learn?

Apply (What was Learned?)
How would you teach someone about this activity or concept?
What did you learn today that you will be able to use in school....at home?
What did you learn by participating in this process that will help you in the future?
Describe a time when you might need the skills/knowledge you learned today.
What are some other situations when you will need to use the skills you learned today?
What would you do differently if you conducted this activity again?

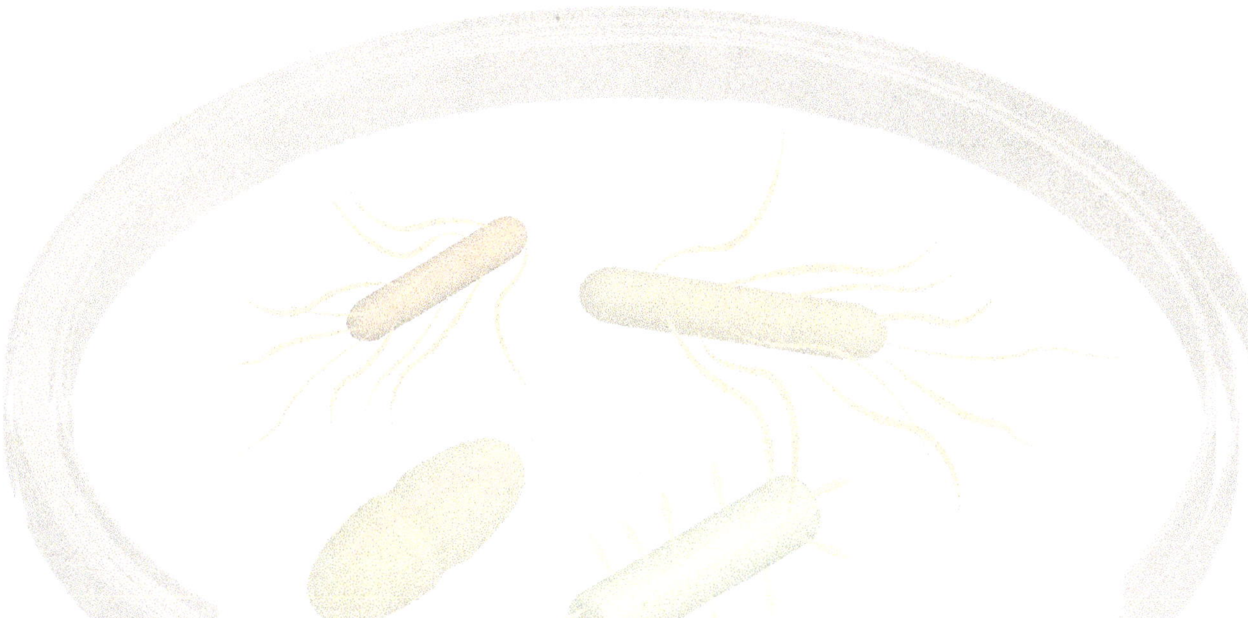

Lesson 1 The Microbes in You:
An Introduction to Microbiology

Introduction

Did you know that there are tiny organisms that are all around you, on you, and inside you? These tiny organisms are called microbes. Microbiology is the study of microorganisms or microbes, which are very small organisms that cannot be seen with the human eye. In today's society, much of the research in the field of microbiology focuses on biochemistry and pathology. Many microorganisms are what are known as pathogens or disease-carrying microorganisms, but not all microbes are pathogens! Some microorganisms are good guys and are necessary in our world. Did you know that the yogurt you eat contains good microbes because yogurt is a live culture? These good microbes help your gut digest food.

Lesson 1 The Microbes in You: An Introduction to Microbiology

Key Teaching Points

1. Microbiology is the study of microorganisms or microbes.
2. Microbes are single-cell living organisms that are too small to see without magnification.
3. Some microbes are pathogens and harmful to our health.
4. Other microbes are beneficial and important to our health.

Learner Outcomes

- Develop a basic understanding of microbiology.

Skill Level: 6-8th grade

Success Indicators

- Participants will identify the basics of microbiology.
- Participants will discuss observations from lab results and what the results mean with regard to daily life.

Life Skills

Teamwork, Healthy living, Critical thinking

Time Needed

2 days, approximately 90-120 minutes in total

Space

Classroom or area where there is ample space to set up desks/seats in two circles

Suggested Group Size

Groups of 4-5 for the lab activity

Materials List

- Technology to present the presentation to the class
- Label Your Microscope worksheet (optional)
- Microbes on My Mind comic (one per participant)
- Microscope Lab - Microbes in Your Mouth instructions (one per group)
- Antiseptic mouthwash
- Microscope (one per group if possible)
- Swabbing materials (per group)
    - cotton swabs
    - 12-14 slides
    - coverslips or additional slides
- Toothbrushes for each participant
- Toothpaste
- Floss for each participant
- Small paper cups (one per participant)
- Computer and internet access for groups
- Recording device for Public Service Announcement (if available, optional; these do not have to be recorded)
- Alcohol swabs to clean the microscope eyepiece

This learning experience will still be effective if participants cannot come to school without brushing their teeth or if this class will be conducted later in the day.

Important Note

Participants should be familiar with the use of a microscope prior to engaging in this lesson. The Microscope Grab and Go is available to train participants and help them become familiar with their microscope and how to view images they may see in this lesson. If you do not have the time available to complete the entire Microscope Grab and Go, the video may be used to introduce the parts and function of the microscope as part of this lesson.

Before the Activity

- The day before this lab, tell participants to come to school without brushing their teeth if at all possible. Tell them that they will be investigating the microorganism jungles that live in their mouths.
- Preview the following videos to determine if you want to include them in the activities for this lesson:

  https://youtu.be/oZ13QfP2os8
  https://youtu.be/zGoBFU1q4g0
  https://youtu.be/hp_8M72V8ew

- Prepare technology to show the presentation to the class.
- Assign participants to groups of 4-5.
- Prepare copies of the Microbes on My Mind comic for each participant.
- Prepare copies of the Label Your Microscope worksheet for each participant if this activity will be completed.
- Prepare copies of the Microscope Lab - Microbes in Your Mouth instructions (one per group).
- Make a decision about the format of the third part of the lesson. Options are given for a whole class discussion or a fishbowl seminar.
- If using the fishbowl seminar, set up desks or seating to accommodate this activity before the class begins if possible.

Fishbowl seminar

1. Create two circles, an inner circle, and an outer circle. The inner-circle should have 6-8 seats or fewer.
2. The facilitator asks open-ended, higher-order thinking Generalize questions.
3. Participants in the outer circle take notes on the participants speaking in the inner circle. They are not allowed to speak or answer any questions.
4. Participants in the inner circle discuss openly the higher-order thinking questions that the facilitator poses.
5. Participants in the outer circle can tap the shoulder of a participant in the inner circle when they have a strong point to make. This is how the circles change and how everyone gets incorporated into the discussion.

6. Circle sizes can be changed to meet the needs of the group, but make sure to have more people in the outer circle than the inner circle.
   It is recommended to have at least 3-4 participants in the inner circle.
7. Facilitators may make it a requirement that everyone must speak X amount of times if desired.
8. For school teachers, there is a rubric that is based on ELA standards to help encourage speaking, writing, and thoughtful reasoning and discussion.

Here is a sample of a fishbowl discussion to use as a reference:
https://youtu.be/ozVc2dDuetc

Opening Questions

- What are microbes? Are they good or bad?
- Why might the study of small microorganisms be important?

Let's Do It!

Optional Part One

View the Microscope Video and have participants complete the Label Your Microscope worksheet.

Part Two

1. Direct participants to move into their small groups.

2. Open the lesson with a challenge to the participants:

   THEY ARE ALIVE AND THEY ARE ALL AROUND YOU!
   Look around the room and tell me if you can name one place where microorganisms are lurking around us. Think of as many as you can in your groups in 2 minutes and then we will share with the class.

3. Ask the Opening Questions.

   - What are **microbes**? Are they good or bad?
   - Why might the study of small microorganisms be important?

4. Direct participants to read the Microbes on My Mind comic with their small groups.

5. Share the presentation with the participants.

6. Have small groups locate the Microscope Lab - Microbes in Your Mouth worksheet. Read the instructions for the lab with the participants as necessary.

Caution: Make sure each participant uses their own swabbing brush for sanitation reasons. Make sure the microscope eyepiece is cleaned with an alcohol swab after each participant uses it.

7. Participants will then discuss the differences they saw from their first slide to their second slide.

8. Ask the Share Questions.

    - What observations did you make about your first slide(s)?
    - What observations did you make about your second slide(s)?
    - What differences did you observe?

9. They will try to answer WHY this occurred and talk about the importance of brushing their teeth. *Try to get this discussion focused on microbes that naturally live in your mouth and the microbes we introduce through everyday activities such as eating food. Participants should begin to realize that when they brush their teeth, they are making their mouth cleaner. Their data/results from their lab should reflect this.*

Optional: If time permits, show one or more of the following short videos about bacteria.

https://youtu.be/oZ13QfP2os8
https://youtu.be/zGoBFU1q4g0
https://youtu.be/hp_8M72V8ew

This is a good stopping point if the class will be divided into two sessions.

Part Three

1. Engage the participants in a whole class discussion or direct the fishbowl seminar.

2. If using the fishbowl seminar, direct participants to seats in the fishbowl.

3. Begin either activity by asking the Reflect question.

    - After looking into the microorganism jungle in our mouths, what do you think is the purpose of antiseptic?
    - Why should we use antiseptic?

4. Continue the chosen discussion by asking the Generalize questions as appropriate.

- What other products around the house fight microorganisms?
- Why is the study of microbiology so important to our daily lives?
- How does understanding microbiology affect your daily life?
- Besides yogurt, what do you think are some other products that contain good microbes?
- Why is it important to have an understanding of good microbes and bad microbes? Give some examples.
- Why do you think it is important to have some good microbes in your mouth? What do you think their role is in your mouth?
- What do humans need to do to protect their bodies from bad microbes or pathogens? Give some specific examples.

5. Direct the participants to rejoin their small groups from the microscope lab.

6. Engage the participants in the Apply activity.

   In their groups of 4-5, participants will create an oral Public Service Announcement about the importance of understanding microbiology in our daily lives. They must specifically mention what are microbes and where they can be found, why individuals personally need to understand good microbes versus bad microbes, and some steps an individual can take to protect their body from bad microbes or pathogens. This should be about 1 minute in length. School teachers can get these Public Service Announcements to be played over the school news and Club leaders can contact local services to see if the participants can get their Public Service Announcement heard throughout the community. These can be recorded with a recording device if it is possible, or a script can be written, at the discretion of the facilitator.

Talk It Over

Share 1	**Reflect 2**
What observations did you make about your first slide(s)? What observations did you make about your second slide(s)? What differences did you observe?	After looking into the microorganism jungle in our mouths, what do you think is the purpose of antiseptic? Why should we use antiseptic?
Generalize 3	**Apply 4**
What other products around the house fight microorganisms? Why is the study of microbiology so important to our daily lives? How does understanding microbiology affect your daily life? Besides yogurt, what do you think are some other products that contain good microbes? Why is it important to have an understanding of good microbes and bad microbes? Give some examples. Why do you think it is important to have some good microbes in your mouth? What do you think their role is in your mouth? What do humans need to do to protect their body from bad microbes or pathogens? Give some specific examples.	In their groups of 4-5, participants will create an oral Public Service Announcement about the importance of understanding microbiology in our daily lives. They must specifically mention what are microbes and where they can be found, why individuals personally need to understand good microbes versus bad microbes, and some steps an individual can take to protect their body from bad microbes or pathogens. This should be about 1 minute in length. School teachers can get these Public Service Announcements to be played over the school news and Club leaders can contact local services to see if the participants can get their Public Service heard throughout the community. These can be recorded with a recording device if it is possible or a script can be written at the discretion of the facilitator.

Lesson 1 The Microbes in You: An Introduction to Microbiology

Term and Concept Discovery

Bacteria: A type of single-celled organism. Bacteria are classified as prokaryotic organisms, as they do not have membrane-enclosed nuclei or other similar structures (organelles). Bacteria are the most abundant life forms on our planet. Singular: bacterium

Microbiology: the branch of biology dealing with the structure,function, uses, and modes of existence of microscopic organisms.

Microbes: any one of a number of organisms too small to be seen without a microscope. Bacteria, viruses, and many fungi are all considered microbes.

Microorganism: Any organism too small to be viewed by the unaided eye, as bacteria, protozoa, and some fungi and algae.

Acknowledgements:
http://webadventures.rice.edu/stu/Games/MedMyst-Original/_301/Game-Overview.html
**Permission has been officially approved from Rice University.

Learn More

http://webadventures.rice.edu/stu/Games/MedMyst-Original/_301/Game-Overview.html

https://asm.org/podcasts

Education Standard(s)

North Carolina 8th grade Essential Standards:
8.L.1-Understand the hazards caused by agents of diseases that affect living organisms.

National Science Standards:
MS-LS1-5-Construct a scientific explanation based on evidence for how environmental and genetic factors influence the growth of organisms.

8th grade English Language Arts Common Core Standards (For optional fishbowl seminar):
8.SL.1-Engage effectively in a range of collaborative discussions (one-on-one, in groups, and teacher-led) with diverse partners on grade 8 topics, texts, and issues, building on others' ideas and expressing their own clearly.

Tags: Microbiology, inquiry learning

Lesson 1   Print Materials

Facilitator Print Materials

Fishbowl Seminar Rubric
Label Your Microscope Answer sheet

Group Handouts

Microscope Lab - Microbes in Your Mouth instructions

Youth Handouts

Label Your Microscope worksheet
Microbes on My Mind Comic

*See page 2 for access to digital assets & extra enrichment.

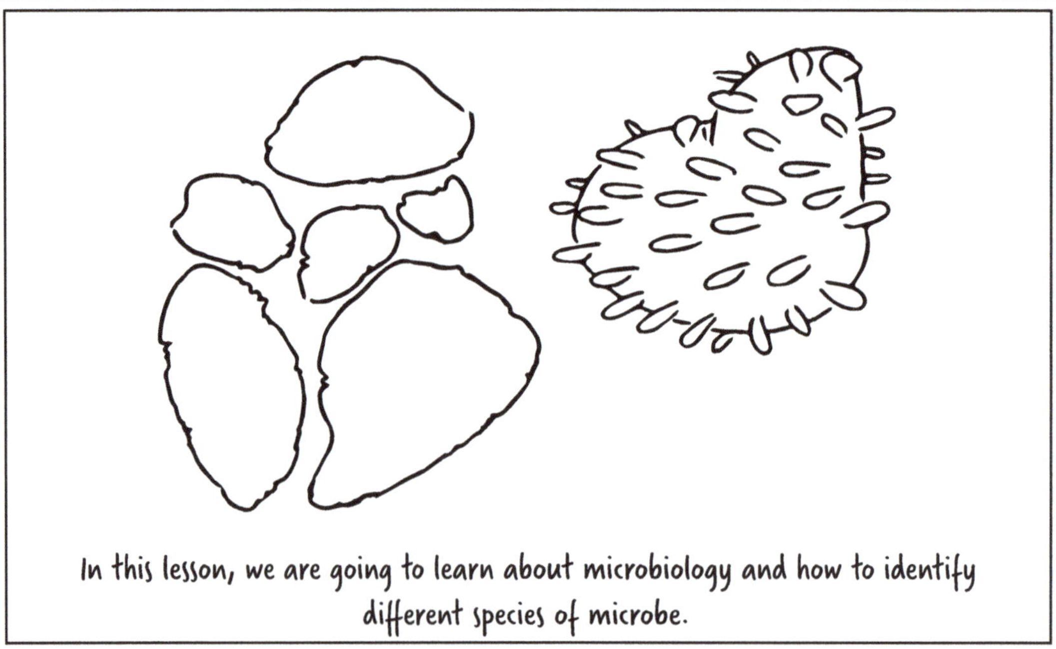

Fishbowl Seminar Rubric

	Excellent 4	Good 3	Fair 2	Unsatisfactory 1
Conduct	Participant: • has patience with different viewpoints of others • asks questions of others • brings other individuals into the conversation • speaks to multiple viewpoints • avoids talking too much	Participant: • generally shows composure and patience throughout the process and responds to different viewpoints • may talk too much or too little	Participant: • somewhat participates and expresses their ideas • does not contribute much to the progression of the conversation	Participant: • does not meet standards for conduct
Speaking & Reasoning (Must speak with true effort at least twice)	Participant: • thoroughly analyzes questions before answering • helps build meaningful discussion and connections and talks about themes • considers others' viewpoints and elaborates or encourages them or contradicts them with research-based evidence	Participant: • responds to some questions • helps engage the conversation to move forward • comments are mostly connected to other ideas	Participant: • responds to questions, but may not add deep meaning to the conversations	Participant: • does not meet standards for speaking and reasoning
Listening	Participant: • pays attention to details • writes down questions and responds to most or all questions • actively listens throughout the process	Participant: • generally pays attention • may only elaborate upon their ideas instead of actively listening to those of other participants	Participant: • may be somewhat inattentive and not actively listening	Participant: • does not meet standards for listening

Fishbowl Seminar Rubric Continued

	Excellent 4	Good 3	Fair 2	Unsatisfactory 1
Reading & Writing	Participant: • is thoroughly familiar with the topic • is fully engaged in the conversation with research-based evidence • takes notes during the seminar • understands key vocabulary words	Participant: • is familiar with the topic • takes some notes during the seminar • understands key vocabulary words	Participant: • is somewhat familiar with the topic • takes few notes during the seminar • has a very basic understanding of the vocabulary words	Participant: • does not meet standards for reading and writing
Overall score: Comments:	/16			

Label Your Microscope Answer sheet

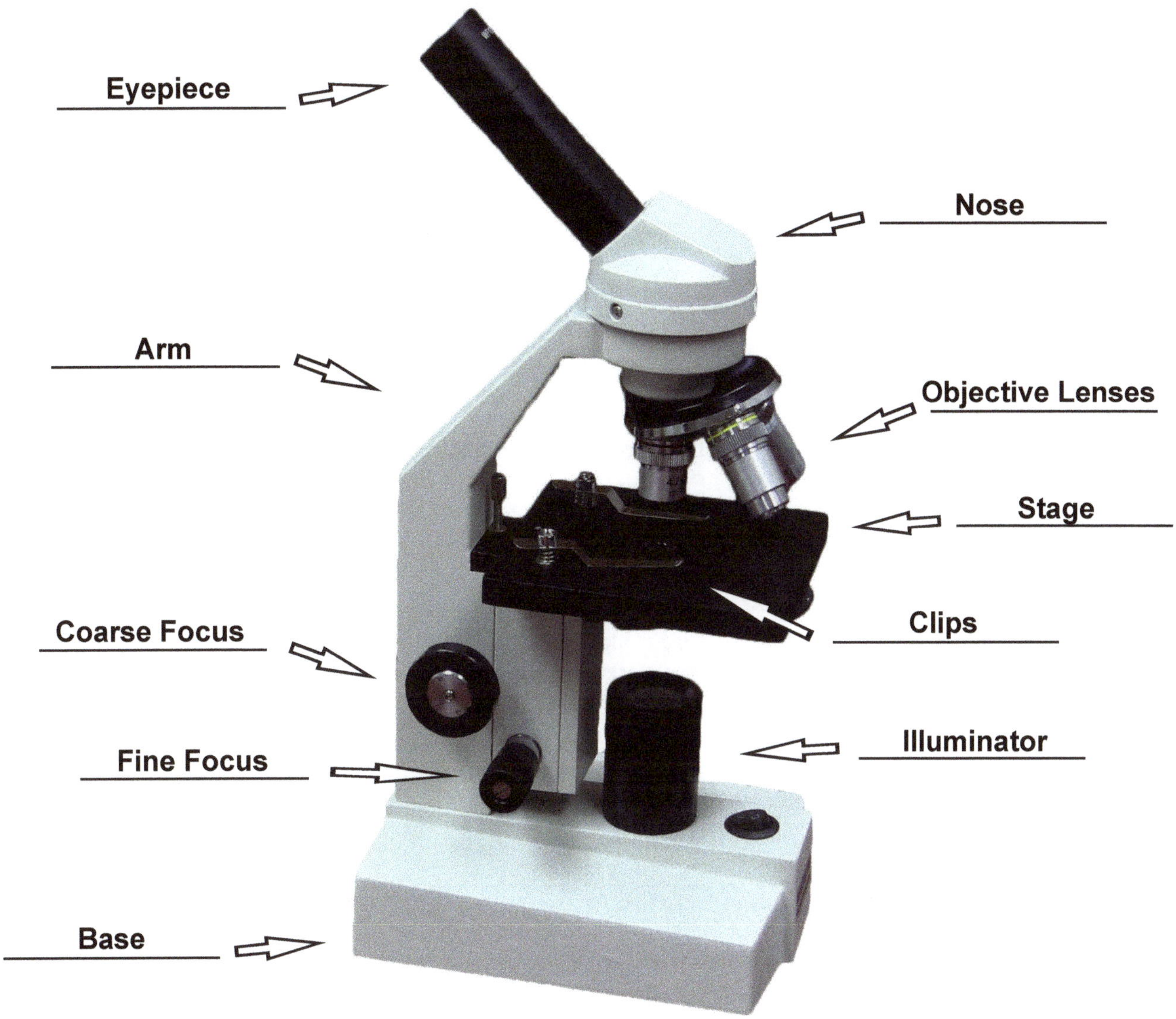

What's my magnification?
Calculate the total magnification of each lens on your microscope below:

______	X	______	=	______
Ocular Lens		Objective Lens		Total Magnification

These answers will depend on the microscopes used in the lesson.

______	X	______	=	______
Ocular Lens		Objective Lens		Total Magnification

______	X	______	=	______
Ocular Lens		Objective Lens		Total Magnification

Microscope Lab - Microbes in Your Mouth

Group names:__________________________________ Date: ____________________________

Materials you will need:

- Microscope
- Swabbing materials:
    - cotton swabs
    - slides
    - coverslips or additional slide
- Toothbrushes
- Toothpaste
- Floss
- Antiseptic mouthwash
- Small paper cups
- Alcohol swabs to clean the eyepiece between each participant

Part One: What's living in your mouth?

Making Your Slide

1. Look at your station. Identify all of the materials you will need for this lab.
2. Identify the parts of the microscope at your station. Operate the moving parts until you are familiar with how they work. Be careful!
3. Select a clean slide. Place it on the workspace in front of you.
4. Select a clean swab.
5. Rub the inside of your cheek and your teeth with the swab. Swab areas where you think the most microbes may live in your mouth.
6. Rub the swab on your clean slide.
7. Cover your slide with a coverslip or another slide.

Viewing Your Slide

1. Rotate the coarse adjustment knob so that the stage moves down and away from the objective lenses as far as it can go.
2. Rotate the nosepiece of the microscope so that the lowest power objective lens is over the iris diaphragm (this should also be the shortest lens).
3. Place your slide on the stage and center the slide over the hole in the stage.
4. Use the stage clips to secure the slide in place. If you are using a second slide as your coverslip and the stage clips disturb your slide, you may decide not to use the clips for this lab.
5. Using the coarse adjustment knob, move the stage all the way up, as close to the objective lens as it will safely move.
6. Look through the eyepiece. Use the coarse adjustment knob and slowly move the stage down until the prepared slide comes into focus.
7. If you move the stage all the way down and do not ever see anything, move your slide on the stage and repeat steps 5 & 6 again.
8. Once the slide comes into coarse focus, use the fine adjustment knob to bring it into sharper focus.
9. Carefully rotate the nosepiece so that the next higher power objective lens is now pointed at your slide. Do not move either the focus knob or the slide during this step.

Microscope Lab - Microbes in Your Mouth Continued

10. Move the fine adjustment knob as necessary to bring the slide into sharp focus again.
11. Sanitize the eyepiece of the microscope with an alcohol swab before the next member of your group uses the microscope.
12. Carefully remove your slide from the stage and clean or dispose of it according to the instructions you are given.

Part Two: How clean is your mouth?

Making Your Slide

1. Brush, floss, and use mouthwash according to your facilitator's instructions.
2. Select a clean slide. Place it on the workspace in front of you.
3. Select a clean swab.
4. Rub the inside of your cheek and your teeth with the swab. Be careful to swab areas where you think the most microbes may live in your mouth.
5. Rub the swab on your clean slide.
6. Cover your slide with a coverslip or another slide.

Viewing Your Slide

Follow the instructions above to use the microscope to view your slide.

Label Your Microscope worksheet

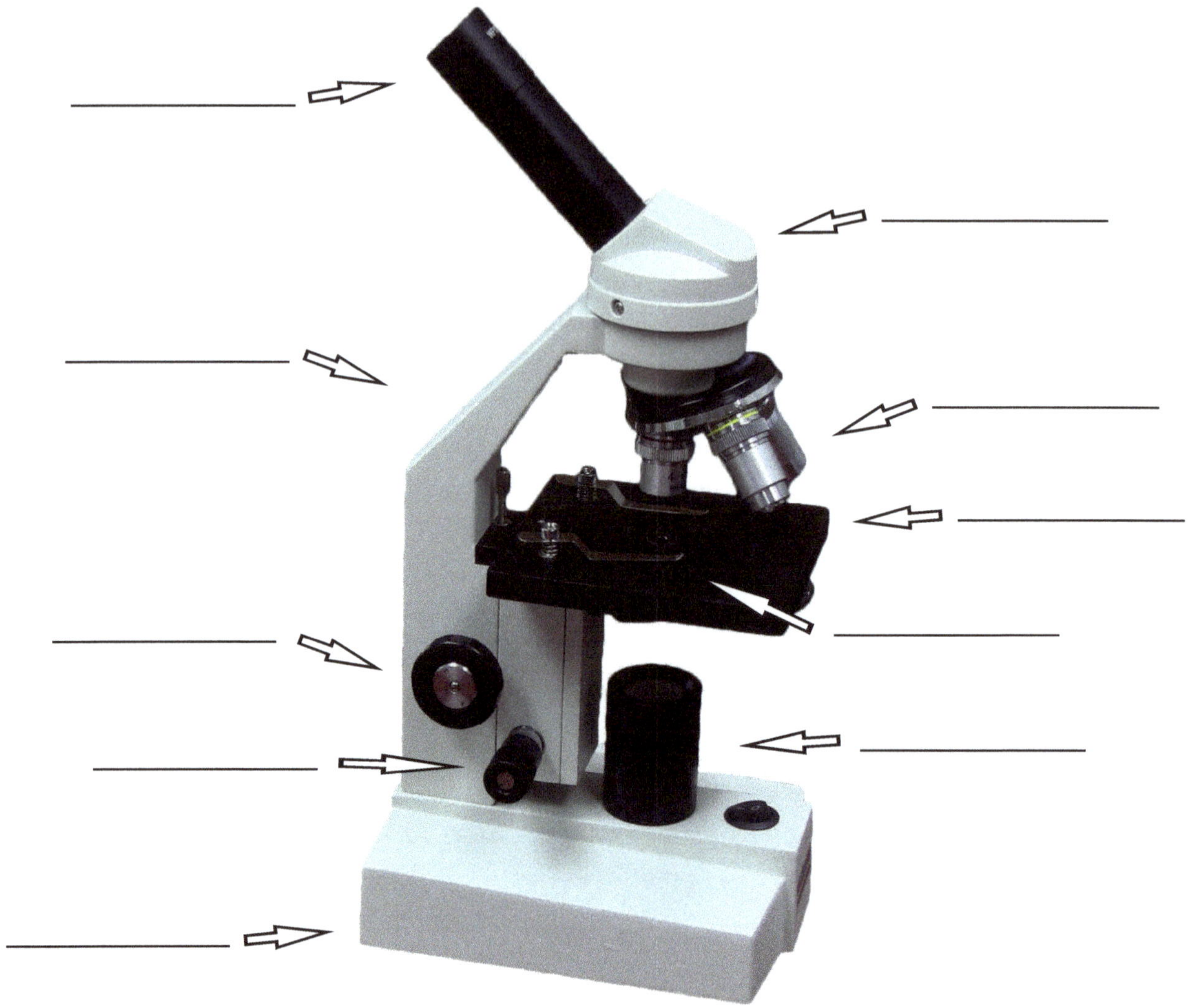

What's my magnification?
Calculate the total magnification of each lens on your microscope below:

______ x ______ = ______
Ocular Lens   Objective Lens   Total Magnification

______ x ______ = ______
Ocular Lens   Objective Lens   Total Magnification

______ x ______ = ______
Ocular Lens   Objective Lens   Total Magnification

Microbes On My Mind Comic

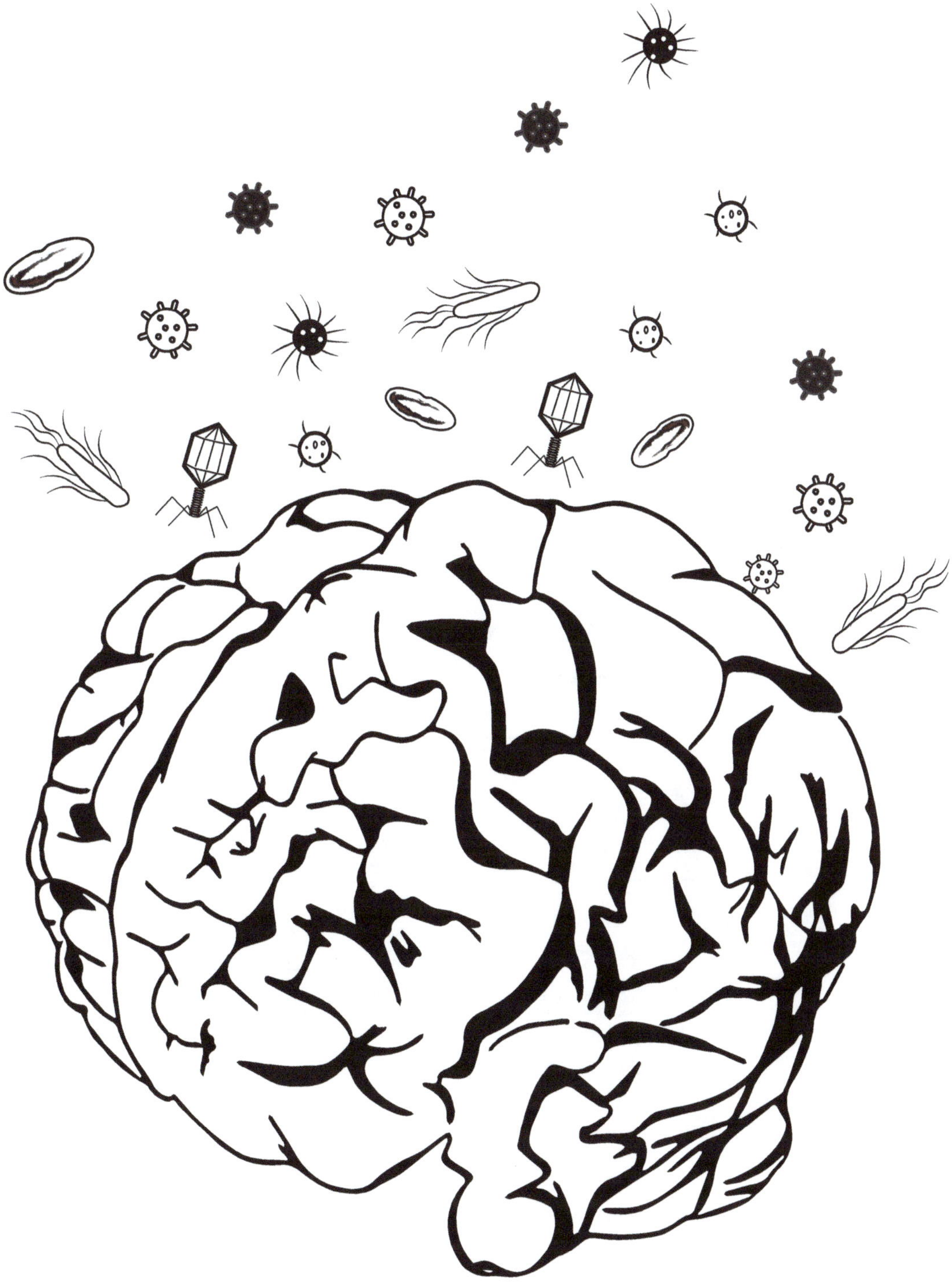

Microbiology is the study of microorganisms or microbes which are very small and cannot be seen with the human eye.

Microbes are the oldest single-celled living organisms on Earth! They are so small that they can fit on the tiny end of a needle.

These microorganisms are small, but tough. They can survive even the harshest conditions, such as extreme heat!

Microbes are literally everywhere! They are in the air, in our food, inside us, and all around us. (The human mouth contains more than 600 types of bacteria.)

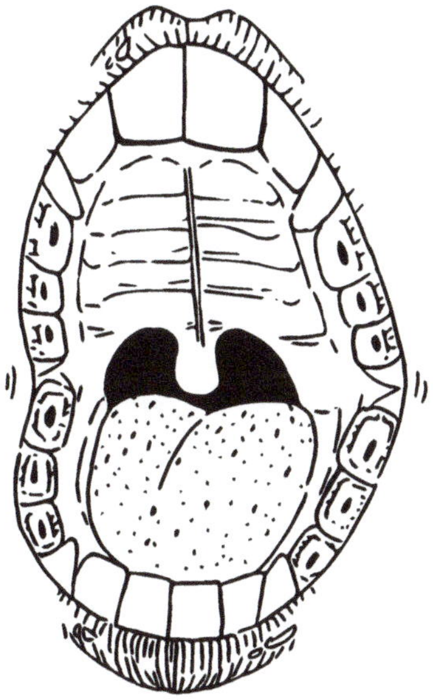

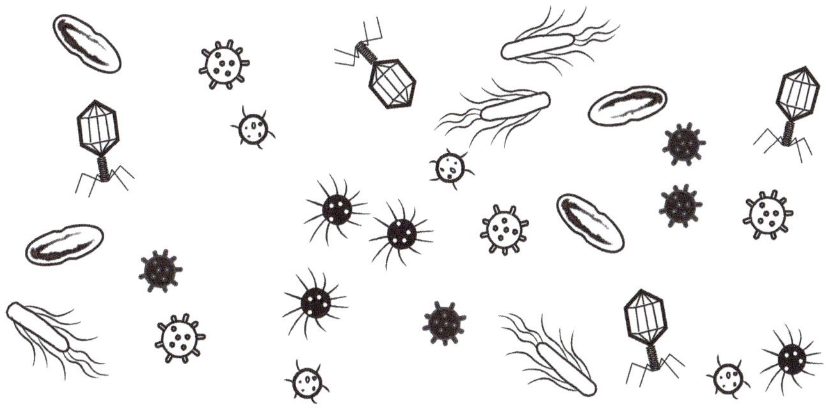

There are good and bad microbes. Not all bacteria are bad, but there are some we do not want in our mouths.

That is why it is recommended to brush your teeth twice a day, floss, and use antiseptic. This practice will help prevent cavities.

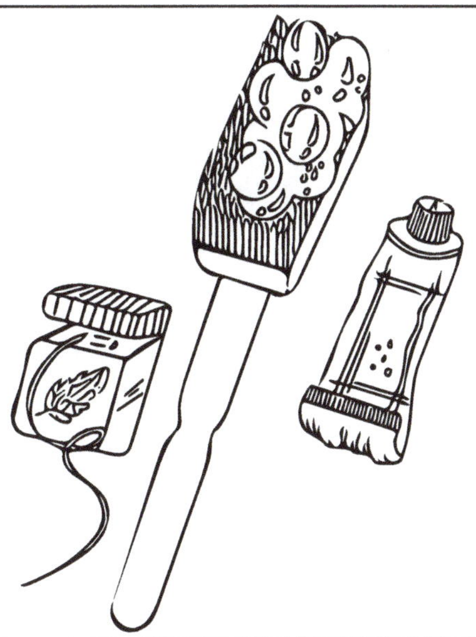

From the beginning of time, microbes have existed. When humans arrived, these microbes had an influence on humans.

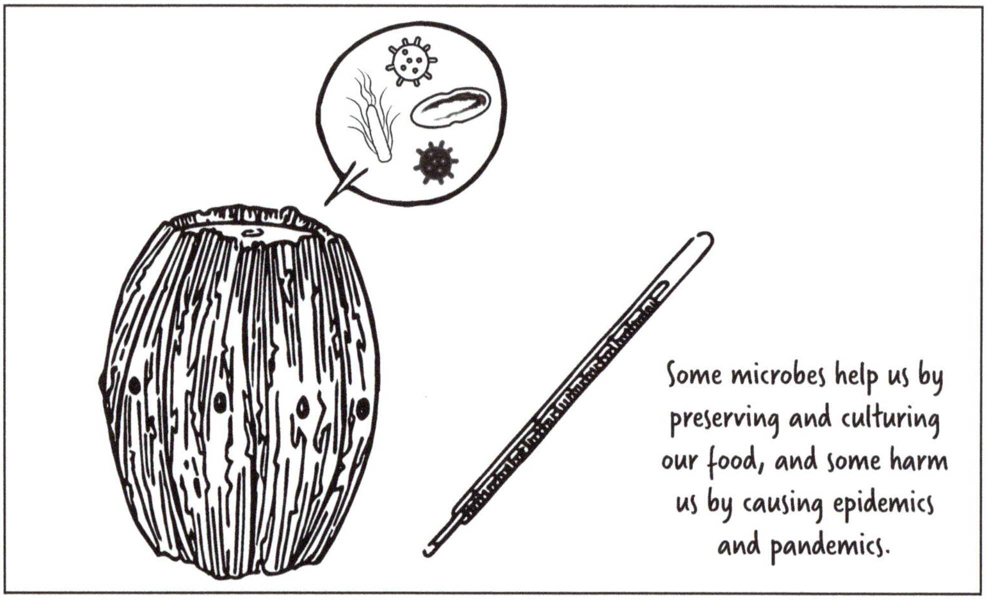
Some microbes help us by preserving and culturing our food, and some harm us by causing epidemics and pandemics.

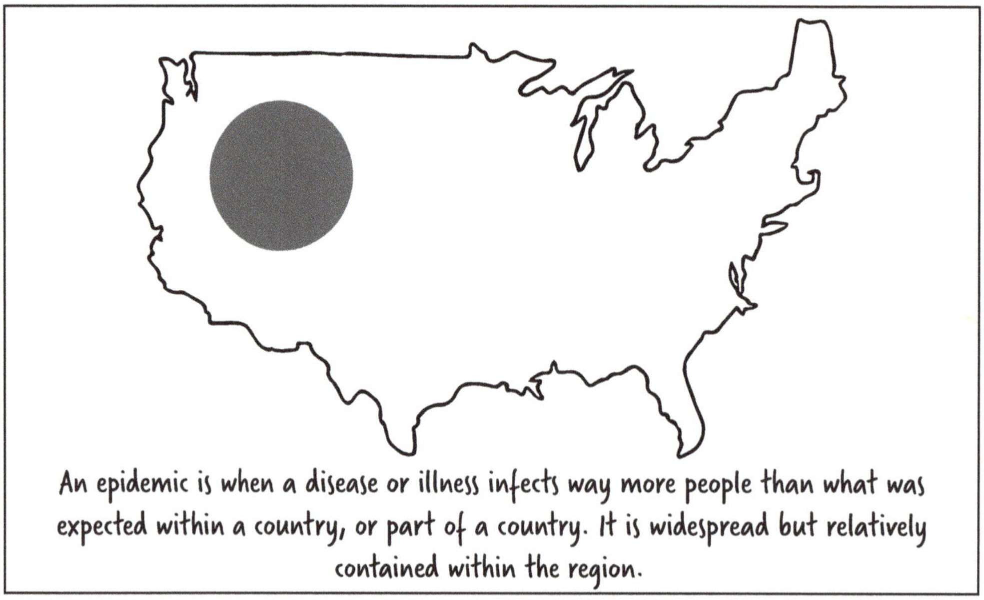
An epidemic is when a disease or illness infects way more people than what was expected within a country, or part of a country. It is widespread but relatively contained within the region.

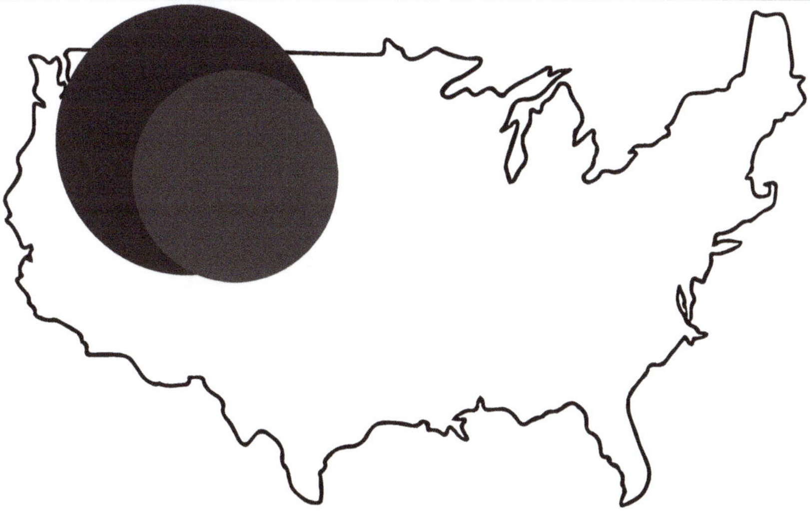

A pandemic is more spread out and geographically severe than an epidemic.

It is usually caused by a new strain of virus or bacteria for which humans have developed little immunity and no cure or medicine.

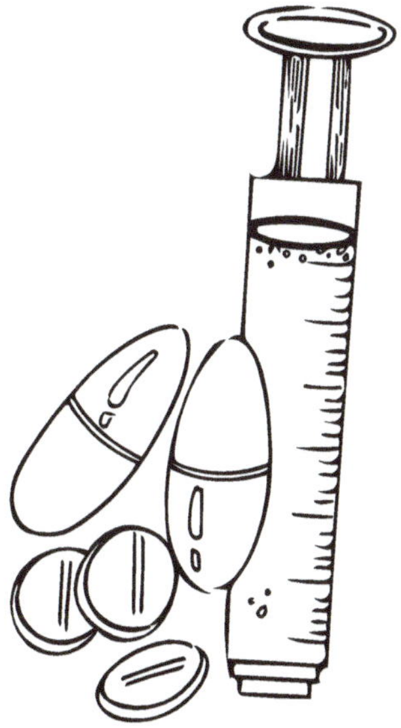

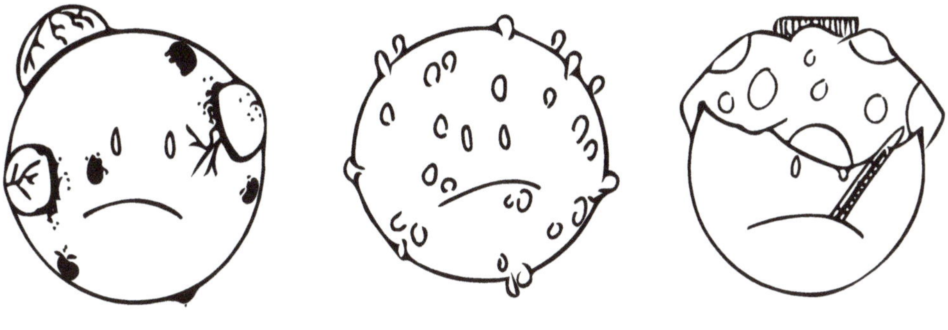

Examples of epidemics and pandemics are: bubonic plague, smallpox, and influenza.

Microbes are in our food. They help us make products such as salami, cheese, and bread.

Did you know, people have been eating yogurt since it was first cultured around 500 BC?

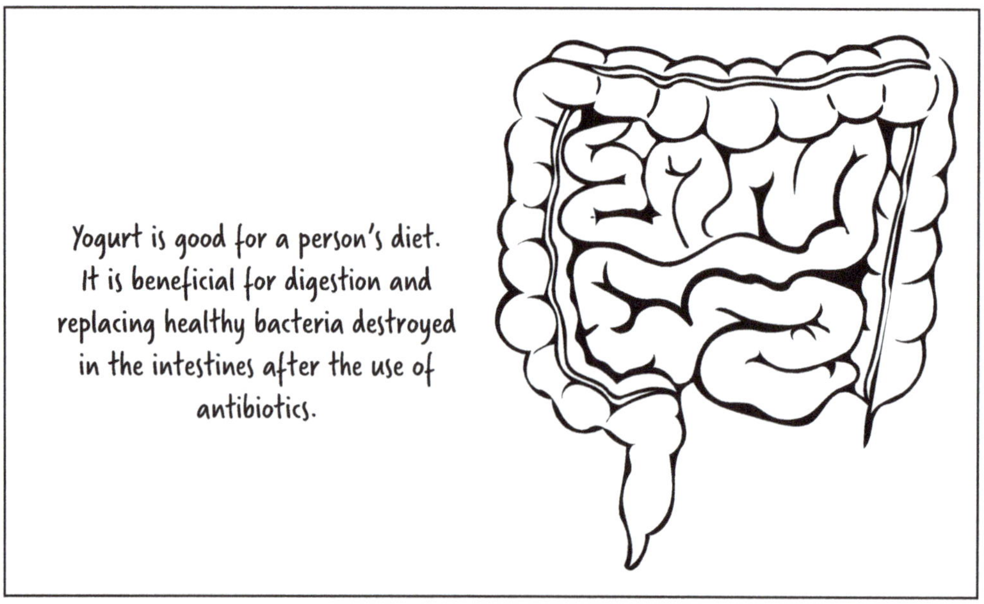
Yogurt is good for a person's diet. It is beneficial for digestion and replacing healthy bacteria destroyed in the intestines after the use of antibiotics.

Just as microbes are in our food, they are also in our water. The types of microorganisms in a water sample tell you a lot about the quality of that water system.

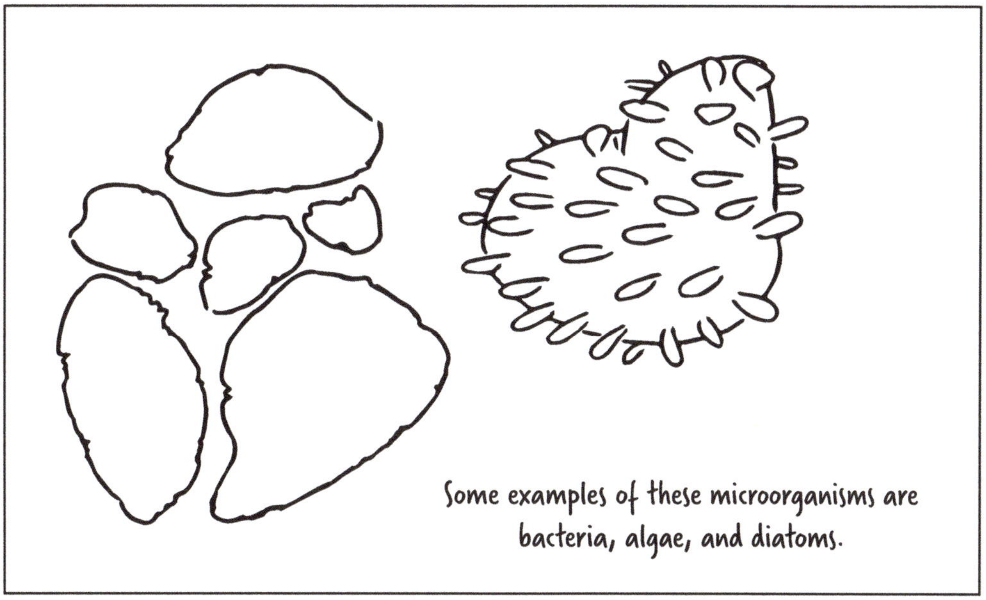
Some examples of these microorganisms are bacteria, algae, and diatoms.

To determine the quantity of microorganisms present in a water system, scientists perform various tests on a water sample. If too many microbes are present, the water is unsafe for consumption.

Microbes are in our food, water, and air. They generate oxygen as a part of the carbon and nitrogen cycles. They also break down matter into nourishing materials for new life to grow.

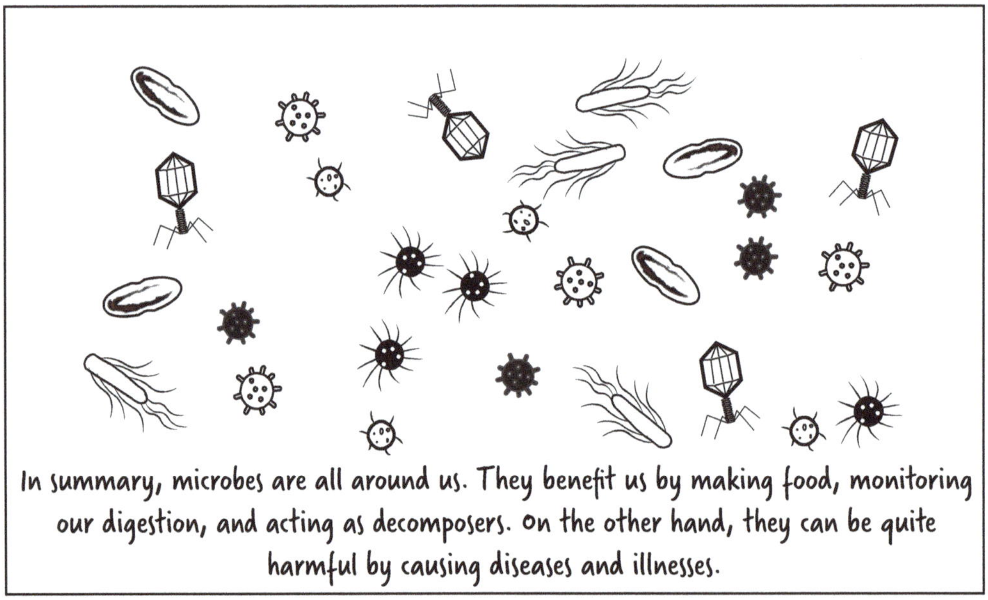
In summary, microbes are all around us. They benefit us by making food, monitoring our digestion, and acting as decomposers. On the other hand, they can be quite harmful by causing diseases and illnesses.

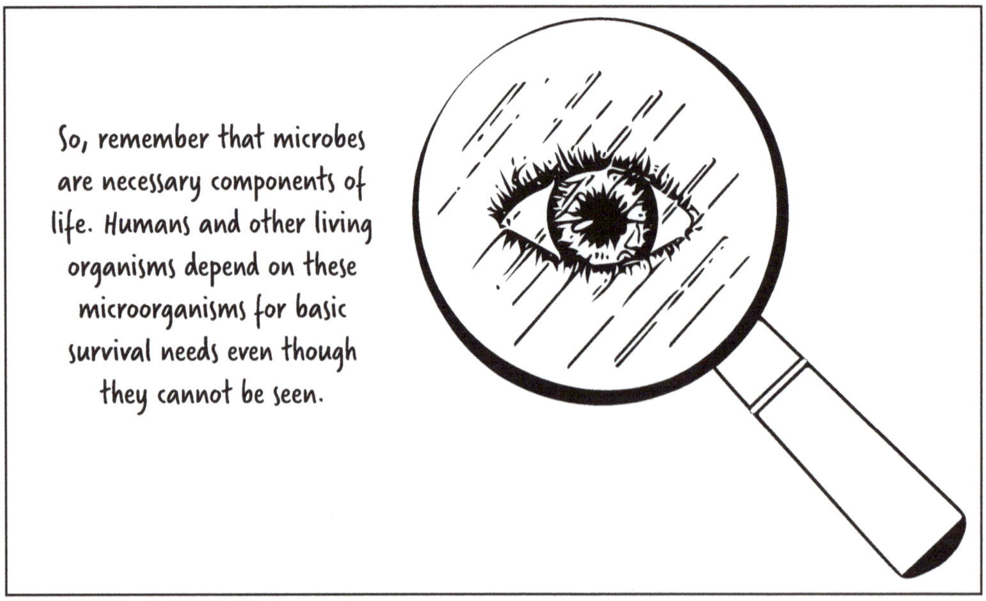
So, remember that microbes are necessary components of life. Humans and other living organisms depend on these microorganisms for basic survival needs even though they cannot be seen.

Lesson 2 Microbes in History

Introduction

Throughout history, many different microbes have plagued our earth. Bad microbes can cause large groups of people to become very sick when an outbreak occurs. When microbes affect huge groups of people, they cause epidemics and pandemics. An epidemic happens when the number of people infected by a microbe rises above the number that was generally expected. For example, if the flu is expected to infect 10,000 people and it infects 15,000 people, this would be an epidemic. When the numbers rise and occur across a larger region of land, such as in several countries, it becomes a pandemic. A pandemic occurs when an epidemic starts infecting even more people and humans have very little immunity to the microbe causing the disease. At this point, it is very easily transmissible or spread from human to human. Pandemics are even worse than epidemics! Many different microbes, such as those that cause influenza (flu) or the bubonic plague (Black Death) have had huge impacts on the world and society.

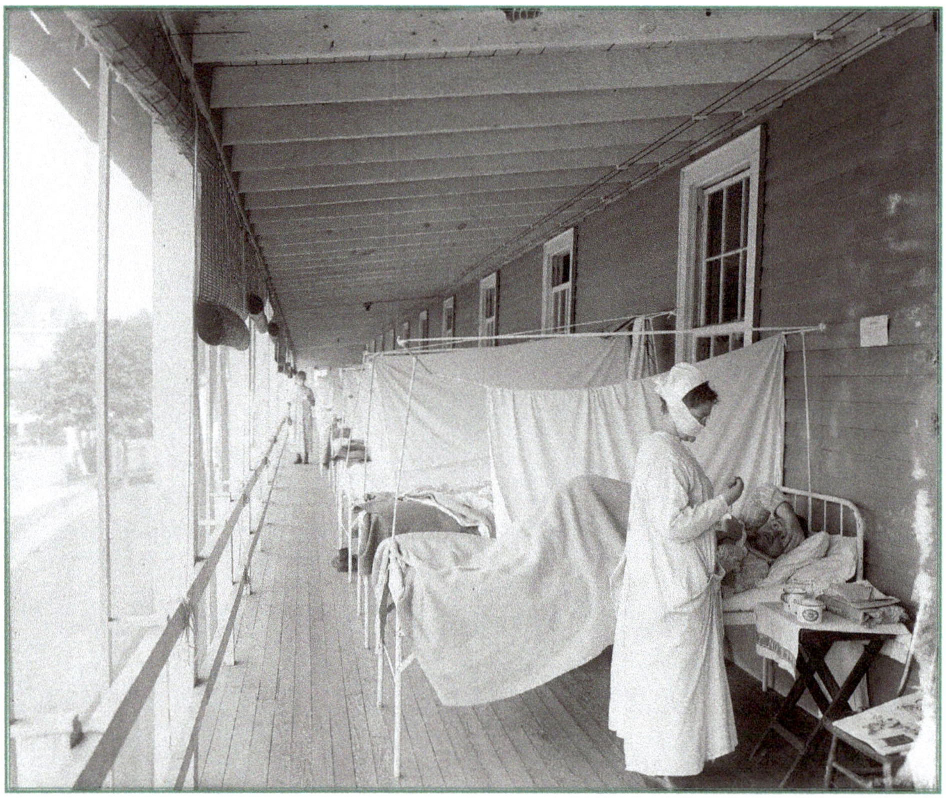

Photo taken during the 1918 Influenza Pandemic, also known as the Spanish Flu, at the Walter Reed Hospital in Washington, D.C.

Lesson 2 Microbes in History

Key Teaching Points

1. Microbes can cause disease in human populations.
2. An epidemic occurs when a disease or illness infects many more people than what was expected within a country or part of a country.
3. A pandemic occurs when a disease or illness infects numerous people over a wider spread geographical area than an epidemic.

Learner Outcomes

- Develop a basic understanding that microbes have plagued human history throughout time.

Skill Level: 6-8th grade

Success Indicators

- Participants will understand the difference between a pandemic and an epidemic.
- Participants will create a wanted Poster for a specific microbe that plagued the human population.

Life Skills

Teamwork, Disease prevention, Critical thinking

Time Needed

2 days, 90-120 minutes in total

Space

Classroom or area where there is ample space for all participants

Suggested Group Size

Participants can create Wanted Posters in groups of 4-5, in pairs, or individually

Materials List

- Technology to present the presentation to the class
- Paper or poster for each participant
- Access to computers or library for participant research
- Glo-germ
- Blacklight
- Colored pencils, markers, rulers, etc., for all participants or groups to complete their Wanted Posters
- Wanted Poster rubric (one per group/participant)
- Think-Pair-Share worksheet (one per participant)

Websites that provide information about diseases, including the CDC, can contain graphic and disturbing images. Please preview the websites that participants will use or provide printed information for their use.

Lesson 2 Microbes in History

Before the Activity

1. Choose one option for participant group size to conduct the Wanted Poster activity. This task can be accomplished by small groups of 4-5, pairs of participants, or individually.
2. Prepare technology to show the presentation to the class.
3. Set up stations for each group or individual to create a Poster.
4. Arrange classroom or group supplies for easy access.
5. Prepare copies of the Wanted Poster rubric for each group/participant.
6. Prepare copies of the Think-Pair-Share worksheet for each participant.
7. Conceal the blacklight somewhere in the room out of sight of the participants.
8. Arrange for technology or print information to be available for participants to conduct research.
9. Immediately before the class, pull three participants aside and have them rub Glo-germ on their hands.
10. Direct all participants to walk back in and act as if nothing has happened. Make sure these particular participants do not share with others that they have the gel on their hands.

Opening Questions

- How do you think germs and diseases are spread?
- What are some ways you think we can prevent diseases from spreading?

Let's Do It!

Part One

1. Ask the Opening Questions.

2. Announce to the group that you want everybody to go around and shake hands with three people. Tell the participants once they have shaken hands with three people, they should sit down.

3. Once all the participants have been seated, tell participants that we have influenza or the flu in our classroom and we had three people infected! We want to see if it has spread to others.

4. Turn off the lights and turn on the blacklight.

5. Each participant will observe their hands underneath the black light.

6. After all participants inspect their hands, have the participants talk about the spread and who shook hands with whom. Have each participant trace from whom they think they were infected.

7. At the end of the activity announce to the class who was infected to see if participants guessed the right people.

    ***Variation to this introduction activity: You can have each participant use an index card to identify with whom they shake hands. After the participants see their hands under the blacklight, they can discuss whose hands they shook and the commonalities.

8. Show the presentation to the class.

9. Introduce the Wanted Poster activity and direct participants either to small groups, pairs, or individual work.

10. Participants will do a quick search on the Internet for facts and pictures of multiple types of microbes in history to gain a basic understanding. This should take no more than 5-10 minutes.

11. After researching, the individual or groups will pick one microbe on which to focus. Each participant or group will create a Wanted Poster for an epidemic or pandemic that has plagued history. Some examples of illnesses caused by microbes are Smallpox, Influenza, The Black Death, Malaria, Tuberculosis, Cholera, Yellow Fever, and Polio.

12. Provide a copy of the Wanted Poster rubric to each participant or group. Instruct the participants or groups to write their name(s) on the rubric.

13. Read the rubric with the participants as necessary before they start. [Note: This rubric is just an example and not required. It may be modified for the facilitator's needs.]

14. Direct participants to use technology or print materials to research and create their Wanted Poster. A time limit may be imposed, and Posters may be completed during the second session if necessary (if the class will be divided). Or, the Poster may be assigned as out-of-class work and brought to the next session.

This is a good stopping point if the class will be divided into two sessions.

Part Two

1. If participants have not finished their research or Posters, allow time at the beginning of this session as necessary.

2. Allow participants to share their Wanted Posters.

3. The class may vote on which microbe is the "Most Wanted."

4. Provide a copy of the Think-Pair-Share worksheet to each participant.

5. Ask the Reflect questions and have participants consider them individually for three minutes.

   - Today, we realize that some microbes are extremely dangerous and can spread during epidemics and pandemics.
   - Why is it important that people study microbes and microorganisms?
   - How did we protect our bodies from microbes in the past versus how we protect our bodies today?
   - What has evolved or changed?

6. Direct participants to discuss their answers with a partner for three minutes and complete their Think-Pair-Share worksheet.

7. Conduct a whole-class discussion to finish the Think-Pair-Share activity.

8. Ask the Generalize questions.

   - Why is it important to study epidemics and pandemics?
   - What difference can humans make in preventing the spread of epidemics and pandemics?

9. Optional: As time allows, allow participants to complete the Apply activity.

   Participants can research an epidemic or pandemic that has affected their region at some point in history. They should look up the statistics, research facts, and create a brochure for their local health department about this microbe that has infected their area.

Lesson 2 Microbes in History

Talk It Over

Share 1

Allow participants to share their Wanted Posters as time permits.

The class may vote on which microbe is the "Most Wanted."

Reflect 2

Conduct a think-pair-share about the following questions:

Today, we realize that some microbes are extremely dangerous and can spread during epidemics and pandemics.

Why is it important that people study microbes and microorganisms?

How did we protect our bodies from microbes in the past versus how we protect our bodies today?

What has evolved or changed?

Generalize 3

Why is it important to study epidemics and pandemics?

What difference can humans make in preventing the spread of epidemics and pandemics?

Apply 4

Participants can do research into an epidemic or pandemic that has affected their region at some point in history.

They should look up the statistics, research facts, and create a brochure for their local health department about this microbe that has infected their area.

Term and Concept Discovery

Epidemic: a disease affecting many persons at the same time, and spreading from person to person in a locality where the disease is not permanently prevalent.

Pandemic: a disease prevalent throughout an entire country, continent, or the whole world; epidemic over a large area.

Lesson 2 Microbes in History

Learn More

Using http://www.cdc.gov/, participants can research the last epidemic or pandemic in recent history or look at current outbreaks (Salmonella is very common) occurring in the United States. Participants can take this information to create a picture book for young participants that explains:

1. epidemics vs. pandemics
2. information about the latest outbreak
3. steps and precautions to take to protect your body from the current outbreak

Participants may view the following video on The Black Death for a deeper understanding of the global impact of the plague.
https://youtu.be/ySCIB6-OH-Q

Significant historical events:
ESRI ArcGIS Blog - https://bit.ly/2TtxgFd
Timelines of History - https://timelines.ws/subjects/Microbiology.HTML

Education Standard(s)

North Carolina 8th grade Essential Standards:
8.L.1-Understand the hazards caused by agents of diseases that affect living organisms.

National Science Standards:
MS-LS1-5-Construct a scientific explanation based on evidence for how environmental and genetic factors influence the growth of organisms.

Tags: history, microbes, pandemic, epidemic

Lesson 2 Print Materials

Facilitator Print Materials

None

Group Handouts

None

Youth Handouts

Wanted poster rubric
Think-Pair-Share worksheet

*See page 2 for access to digital assets & extra enrichment.

Wanted Poster rubric

Name(s): ______________________________ Date: ________________________

Create a Wanted Poster for your pathogen. Some information to include:

- The name of the disease
- The name of the microbe
- A picture of the microbe
- Why the microbe is wanted (information about the epidemic or pandemic)
- How it is spread
- Symptoms of the disease
- When and where it happened
- How many people were killed
- How to protect yourself from the microbe

Be colorful, be creative, have fun!!

Criteria	3	2	1	0
Research	Pathogen has been fully researched and includes 6-7 relevant facts about the epidemic or pandemic.	Pathogen has been researched and includes 4-5 facts about the epidemic or pandemic.	Pathogen has been researched and includes 2-3 facts about the epidemic or pandemic.	Research was not used to guide the design process of the poster.
Information about Epidemic/ Pandemic	The poster includes very interesting and accurate information and accurately explains why the microbe is wanted.	The poster includes interesting and accurate information and explains why the microbe is wanted.	The poster includes accurate information and somewhat explains why the microbe is wanted.	The poster includes inaccurate information and/or does not explain why the microbe is wanted.
Colorful, Appealing	The poster is very colorful and appealing to the eye.	The poster is colorful and appealing to the eye.	The poster is somewhat colorful and appealing to the eye.	The poster is not as appealing as it could be.
Organization and Creativity for All Aspects	The group has an extremely neat and orderly poster.	The group has a neat and orderly poster.	The group has a somewhat neat and organized poster.	Organization or neatness is not evident in the poster.
Total Points: /12 Comments:				

Think-Pair-Share worksheet

Name(s): ______________________________ Date: ______________________

THINK (What I Think)	PAIR (What You Think)	SHARE (What We Think)

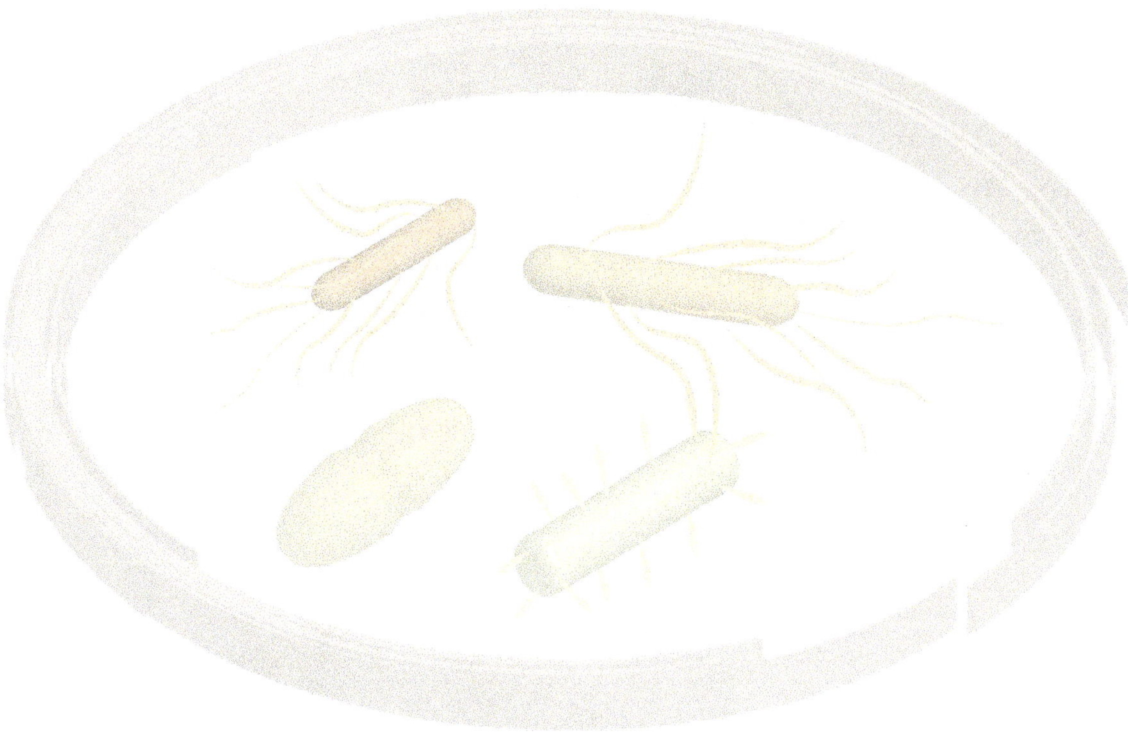

Lesson 3 Microbes in Our Food

Introduction

Believe it or not, microbes are found in many of the foods we consume daily. Some of the foods we consume would not develop if it wasn't for these microbes. One example is yogurt. The process of making yogurt has been around for centuries. In this activity, participants will learn about the microbes (bacteria) that are used to create yogurt as well as learn how to make their own culture of yogurt.

Lesson 3 Microbes in Our Food

Key Teaching Points

1. Microbes are found in many of the foods we consume on a daily basis.
2. Yogurt and some other foods we consume would not develop if it wasn't for these microbes.

Learner Outcomes

- Understand the roles of microorganisms in food and the importance of food safety.

Skill Level: 6-8th grade

Success Indicators

- Participants will understand the importance of food safety.
- Participants will successfully make yogurt.
- Participants will create an advertisement promoting the yogurt they created and the importance of food safety.

Life Skills

Teamwork, Critical Thinking, Communication, Contributions to group efforts, Marketable Skills, Disease Prevention, Problem Solving

Time Needed

1-2 days

Space

Classroom or area where participants can work in groups

Suggested Group Size

3-4 participants per group

Materials List

- Technology to present the presentation to the class
- Microbes in Our Food Introduction and Lab Analysis worksheets (1 per participant)
- Microbes in Our Food Lab Procedure (1 per group)
- Plain yogurt (at least 1 cup is necessary for making yogurt to use as a starter)
- Vanilla or other flavored yogurt to taste (optional - enough for each participant)
    - **Note:** Yogurt for tasting can be any flavor. The starter yogurt needs to be plain yogurt.
- Styrofoam cups with lids (1 per group)
- Plastic spoons (1 per person for yogurt bar and 2 per group when making yogurt)
- Quart sized sealable bags (1 per group)
- Plastic wrap (1 roll)
- Newspaper (any color works)
- Napkins (enough for all participants)
- Paper cups for yogurt tasting (1 per participant)
- Whole milk (1 gallon)
- Powdered milk (1 box)
- Thermometer (digital works best)
- Permanent marker (1 per group)
- 1 hot plate
- 1 large (2 qt. or larger) beaker or saucepan
- 1 oven mitt for holding sauce pan
- 1 cooler or insulated container to insulate yogurt cultures
- Toppings for yogurt tasting (these can be provided by facilitator or participants)

Explain to participants that they will NOT be tasting or eating the yogurt that they make in this learning activity. The yogurt participants may taste is commercially prepared.

Lesson 3 Microbes in Our Food

Before the Activity

Up to a week prior to the activity:

1. Invite participants to bring in toppings for the activity for the day of the lesson. Do not tell participants that they will be making yogurt. Explain to participants they want to bring in toppings similar to ice cream toppings.
2. If the facilitator prefers to supply toppings, consider buying items like cut-up fruits, granola, flavored syrups (chocolate, strawberry), and other toppings that participants might enjoy.

   *__Note:__ be aware of any food allergies (nuts, dairy products, etc.) participants may have

Day of activity:

1. Prepare technology to show the presentation to the class.
2. Prepare copies of the Microbes in Our Food Introduction and Lab Analysis worksheets (one per participant).
3. Prepare copies of the Microbes in Our Food Lab Procedure (one per group).
4. Bring in different items or provide pictures of different items that go through the fermentation process (bread, sauerkraut, soy sauce, cheese, etc.).
5. Have a "yogurt bar" set up for participants before they enter the room.
6. Set up the materials for each group:
   - 1 foam cup with lid
   - 1 teaspoon of starter yogurt
   - 1 plastic spoon
   - 1-quart size sealable bag
   - Newspaper
   - Plastic wrap
   - 1 permanent marker
   - Microbes in Our Food Lab Procedure

****CAUTION: for health and safety purposes participants should not eat the yogurt they create. This activity is to give them an understanding of how bacteria help in the production of yogurt.**

Opening Questions

- How do you think yogurt is made?
- Where do you think the concept of making yogurt originated from?
- Describe the texture and taste of your yogurt.
- How do you think different flavors of yogurt are made?

Lesson 3 Microbes in Our Food

Let's Do It!

1. Give participants a small cup with yogurt and allow them to place toppings of their choice on it.

2. Provide each participant with one copy of the Microbes in Our Food Introduction and Lab Analysis worksheets.

3. When all participants have their samples and are enjoying their treat, ask the Opening Questions.

   - How do you think yogurt is made?
   - Where do you think the concept of making yogurt originated?
   - Describe the texture and taste of your yogurt.
   - How do you think different flavors of yogurt are made?

4. Direct participants to take notes from the class discussion for each question on the worksheet.

5. Engage participants' questions/responses about the production of yogurt. Ask participants to come up with a list of ideas they have about what is required to make yogurt.

6. Show participants the presentation on how yogurt is made and the importance of fermentation. Direct them to fill in the missing words on their worksheets.

7. Show participants any items or pictures of different items that go through the fermentation process (bread, sauerkraut, soy sauce, etc.) that you brought in and discuss with participants the importance of fermentation and bacteria in food production.

8. Direct participants to join their small groups.

Facilitator Lab Procedure

This first part of the activity will be conducted by the facilitator. It is highly suggested participants watch this procedure being done. This gives the facilitator a chance to discuss why it is important for the mixture to be heated before the production of yogurt. (Bacteria thrive in warm and wet environments, and we want the bacteria in the yogurt to grow.) This is also an opportunity for the facilitator to discuss why certain foods have to be cooked before consumption. This connects to the food safety aspect of this lesson.

The following mixture will make enough for 4 groups. Double the amounts as needed for more groups.

1. Plug in the hot plate and set to medium-high.
2. Add 3-3 ½ cups whole milk to a large beaker or saucepan and place it on the hot plate.
3. Add ¾ cup of powdered milk into the liquid milk.

4. Stir the mixture slowly and constantly.
5. Use the cooking thermometer to monitor the temperature of the milk.
6. When milk gets between 175-200ºF, remove from heat. (**Note**: yogurt will not set up properly if the milk boils.)
7. Cover the top of the milk with plastic wrap to prevent a film from developing on top of the substance.

8. Allow the milk to cool to approximately 135ºF.
9. Once the milk has reached the desired temperature, remove the plastic wrap.
10. Dispense the milk into four foam cups (one cup per group).

Participant Lab Procedure: Part 1

Participants will need to have all materials listed on the Lab Procedure at their workstations.

1. Label the bag using a permanent marker with group members' names.
2. Use the spoon to mix the milk to make sure nothing has started to settle.
3. Add one teaspoon of yogurt starter into the cup.

4. Stir the mixture carefully but make sure the starter gets evenly distributed throughout the milk.
5. Once the solution is completely mixed, cover the cup with plastic wrap.
6. Place the plastic lid on top of the cup over the plastic wrap.

7. Carefully place the cup down into the sealable bag.

8. Take the newspaper provided and rip it into shreds. Place the newspaper shreds around the cup in the bag to help retain heat.

9. Make sure the bag is tightly sealed. The top of the bag may also be taped to help prevent spills.

10. Place the yogurt mixture into an insulated container for 6 hours or longer (overnight is acceptable).

Participant Lab Procedure: Part 2

1. Get the yogurt sample from the insulated container.
2. Get a clean plastic spoon from the facilitator.
3. Carefully remove the cup from the baggie.
4. Carefully remove the lid and the plastic wrap from the cup.
5. Observe the yogurt. Use the clean spoon to move yogurt around to look at the consistency.
6. Compare the group's yogurt sample with the other groups' samples.
7. Discuss, as a group, the role bacteria played in making the yogurt. Discuss why the milk went from a liquid to more of a solid and how the bacteria in the yogurt contributed to it.

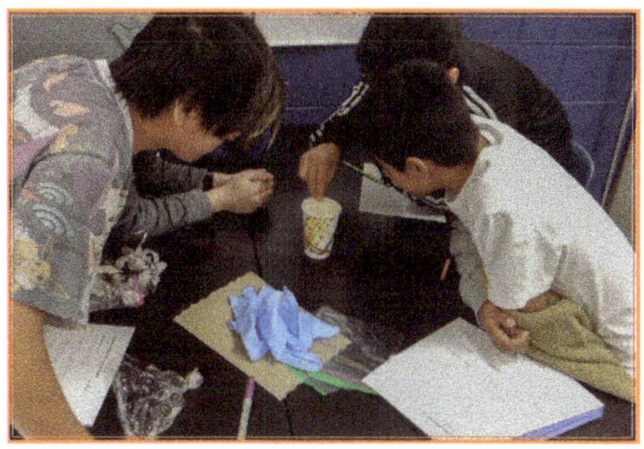

8. Direct participants to complete the Share, Reflect, Generalize, and Apply questions on their Lab Analysis worksheet with their small groups.
9. When the groups are done observing the yogurt, carefully dispose of the cups into the trash can.
10. Make sure workstations are sanitized using proper cleaning materials.

Talk It Over

Share 1	Reflect 2
Compare the yogurt you made to the yogurt you tasted at the beginning of the activity. Discuss color, smell, and consistency. How did your yogurt sample compare to other groups' samples? Why do you think other samples may have turned out differently? Why do you think store bought yogurt must be refrigerated?	How were bacteria useful in the production of your yogurt? Why do you think fermentation is an important process in the production of certain foods? What other foods do you know go through a fermentation process?
Generalize 3	**Apply 4**
What general statements could you make about the importance of microorganisms in our food?	How would you convince your friend that some bacteria actually are good for you? What are some examples you would use?

Term and Concept Discovery

Bacteria: a type of single-celled organism. Bacteria are classified as prokaryotic organisms, as they do not have membrane-enclosed nuclei or other similar structures (organelles). Bacteria are the most abundant life forms on our planet. Singular: bacterium

Fermentation: a change brought about by a ferment, as yeast enzymes, which convert grape sugar into ethyl alcohol.

Food Safety: a series of practices designed to minimize risks posed by microbes in the food supply. Food safety practices include proper hand washing, cooking food thoroughly, keeping perishable items chilled, cleaning food preparation surfaces and tools, and keeping different types of food separated to avoid cross-contamination.

Microorganism: any organism too small to be viewed by the unaided eye, as bacteria, protozoa, and some fungi and algae.

Probiotic: a usually dairy food or a dietary supplement containing live bacteria that replace or add to the beneficial bacteria normally present in the gastrointestinal tract.

Variations and Extensions

1. With a group create an advertisement promoting a brand of yogurt you created. Include information about why your yogurt should be bought over others. Discuss the benefits of eating yogurt and why your yogurt is considered to be a healthy choice for consumers.
2. Spend time researching more about how certain foods are produced through fermentation. Create a presentation on the importance of fermentation in food production.
3. Participants may want to create a bacterial culture with their yogurt to observe growth. Use a sterile petri dish with agar to conduct this activity. Participants can also place a small amount of the yogurt under a microscope to observe what the magnified yogurt sample looks like.

If participants can't make yogurt, the facilitator can prepare some before the meeting and show participants the process and the product of making yogurt.

References

https://www.culturesforhealth.com/learn/yogurt/what-is-yogurt-history/

http://www.explorebiology.com/documents/LE/LabMakingYogurt2008.pdf

Source and more information of the history of bacteria:

Learn More https://www.culturesforhealth.com/learn/yogurt/what-is-yogurt-history/

Education Standard(s)

North Carolina 8th grade Essential Standards:
8.L.2:Understand how biotechnology is used to affect living organisms

National Science Standards:
MS-LS1-1: Conduct an investigation to provide evidence that living things are made of cells; either one cell or many different number and types of cells
MS-LS1-5: Construct a scientific explanation based on evidence for how environmental and genetic factors influence the growth of organisms

Tags: microbiology, food safety, inquiry learning, fermentation, bacteria, microorganism

Lesson 3 Print Materials

Facilitator Print Materials

None

Group Handouts

Microbes in Our Food Lab Procedure

Youth Handouts

Microbes in Our Food Introduction

Lab Analysis worksheets

*See page 2 for access to digital assets & extra enrichment.

Microbes in Our Food Lab Procedure

1 Styrofoam cup with lid 1 teaspoon of starter yogurt 1 plastic spoon (Part 1) 1 quart size zip-lock bag	Newspaper Plastic wrap 1 permanent marker 1 plastic spoon (Part 2)

Part 1 - Procedures

The beginning procedure for this lab will be conducted by the facilitator, who will then distribute cups of warm milk to each group.

1. Label the bag using a permanent marker with group members' names.
2. Use the spoon to mix the milk to make sure nothing has started to settle.
3. Add one teaspoon of yogurt starter into the cup.
4. Stir the mixture carefully but make sure the starter gets mixed throughout the milk.
5. Once the solution is completely mixed, cover the cup with plastic wrap.
6. Place the plastic lid on top of the cup over the plastic wrap.
7. Carefully place the cup down into the sealable bag.
8. Take the newspaper provided and rip it into shreds. Place the newspaper shreds around the cup in the bag to help retain heat.
9. Make sure the bag is tightly sealed. The top of the bag may also be taped to help prevent spills.
10. Place the yogurt mixture into an insulated container for 6 hours or longer.

Part 2 - Procedures

1. Get the yogurt sample from the insulated container.
2. Get a clean spoon from the facilitator.
3. Carefully remove the cup from the baggie.
4. Carefully remove the lid and the plastic wrap from the cup.
5. Observe the yogurt. Use the clean spoon to move yogurt around to look at the consistency.
6. Compare your yogurt with the other groups' yogurt.
7. Discuss, as a group, the role bacteria played in making the yogurt. Discuss why the milk went from a liquid to more of a solid and how the bacteria in the yogurt contributed to it.
8. Complete the questions on your Lab Analysis worksheet with your small group.
9. When the group is done observing the yogurt, carefully dispose of the cup into the trash can. Make sure the work area is clean.

Microbes in Our Food Introduction

Name(s): ______________________________ Date: ________________________

Pre-Lab Questions

How do you think yogurt is made?

Where do you think the concept of making yogurt originated?

Describe the texture and taste of your yogurt.

How do you think different flavors of yogurt are made?

Brief History of Yogurt

Yogurt is a ______________ product that originated in ______________. The word ____________ is Turkish for "curdle" or "____ __________". It is possible that yogurt was discovered by ____________ when it was first produced around ___________ BC. ___________ would store the milk of cows in bags that were made from the _____________ of animals. The combination of the warm environment and the ___________ found in these bags made the milk curdle and become more ___________. People began to eat this interesting product. The acidity of the ___________ also allowed it to keep longer and stay ______________. It quickly became a popular food in the ______________ ______________ and eventually spread ______________.

Yogurt can be very beneficial to a person's _________. Some of the bacteria that is found in yogurt is very similar to the bacteria found in the _____________. When people take __________________, not only does it kill the bad bacteria, but it can also kill off some of the ____________ ____________ that help with ____________. This is why it is recommended to eat yogurt with an antibiotic because this helps in replacing the good bacteria in the digestive tract. Yogurt helps in digestion and helps keep the __________ of good bacteria up in the body.

Lab Analysis worksheet

Part 1 Lab Observations

1. Describe what your yogurt looks like on Day 1.

2. Draw what your yogurt looks like on Day 1.

3. Predict what you think your yogurt will look like on Day 2 and state why.

Lab Analysis worksheet Continued

Part 2 Lab Analysis

1. How did your yogurt sample compare to other groups' samples? Why do you think other samples may have turned out differently?

2. Why do you think store bought yogurt must be refrigerated?

3. How were bacteria useful in the production of your yogurt?

4. Why do you think fermentation is an important process in the production of certain foods?

5. What other foods do you know that go through a fermentation process?

6. What general statements could you make about the importance of microorganisms in our food?

7. How would you convince your friends that some bacteria is actually good for you? What are some examples you would use?

Lesson 4 Microbes in Our Water

Introduction

Microbes are found all over our world, including in the water. The microbes found in water tell us a great deal about the quality of the water and the organisms that live in it. In some cases, a water system is considered healthy if diverse groups of microbes can be found living in it. On the other hand, sometimes the presence of a particular microbe can indicate that this water system is not a healthy body of water for certain organisms to call home. In this lesson, participants will explore the different types of microorganisms found in water samples and compare their findings to other samples of water.

Untreated water specimens from a stream containing unidentified organisms. phil.cdc.gov

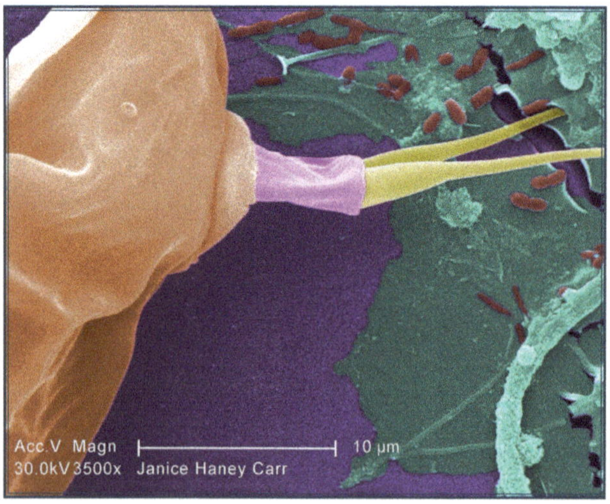

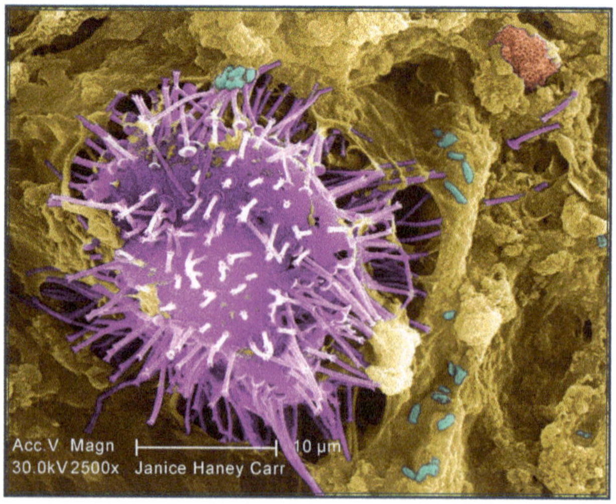

Lesson 4 Microbes in Our Water

Key Teaching Points

1. A variety of microbes are found in water sources.
2. The type of water environment will determine the variety of microorganisms found there.

Learner Outcomes

- Identify microorganisms in water sources.

Skill Level: 6-8th grade

Success Indicators

- Participants will collect water samples.
- Participants will identify microorganisms found in the water sample.
- Participants will gain an understanding of the diversity of microorganisms found in water sources.
- Participants will compare water samples and the diversity of microorganisms from different water samples.

Life Skills

Teamwork, Critical Thinking, Communications, Contributions to Group Efforts, Marketable Skills, Disease Prevention, Problem Solving

Time Needed

1-2 days

Space

- Outdoor area with water source (pond, lake, stream) to collect samples
- Classroom space to complete the testing and identifying of microbes

Suggested Group Size

4-5 participants in a group

Materials List

For the class

- Technology to present the presentation to the class
- Microbe Data Sheet (1 per participant)
- 1 empty clear jar (mason or baby food jars work well)
- Sample of water from a local water source (pond, river, stream)
  Note: Try to get a sample that looks to be "clear with nothing in it"

For Each Group

- 1 empty baby food jar or other small clear jar with lid (cleaned and label removed)
- Microscope Lab Procedures (1 per group)
- Hand lens
- Microscope (if available)
  - Microscope slides
  - Coverslips for microscope slides (one slip needed per slide). [Optional: additional slides may be used in place of coverslips.)
  - Pipette (dropper)
- Alcohol swabs to clean the eyepiece between each participant
- Slide cleaning materials to use after the lab if the slides will be reused
  - **Options:** soap and water, glass cleaner, alcohol spray, or lens cleaner AND paper towels, tissues, or lint-free tissues
- **Optional:** Water Testing kit can be used to expand discovery of the water sample

This learning activity can be completed with or without microscopes, depending on what resources are available.

Before the Activity

Up to a week before the activity:

- Determine from where participants will collect water samples. One water source may be used, or each group may collect water from a different source.
- Inform the participants and their parents or guardians that they will be outside collecting water samples on the scheduled date. Collect permissions as necessary.

The day before the activity:

- Make sure the weather will permit outside activity.
- Remind participants that they will be outside the following day and to dress accordingly.
- Collect a sample of water from a local water source (pond, river, stream). This can be one jar for the whole group or multiple to pass around. Make sure the sample is in a clear container (Mason jar, baby food jar). Try to get a sample that is fairly clear and looks as if nothing is present in the water sample.

Day of activity:

- Prepare technology to show the presentation to the class.
- Prepare copies of the Microbe Data Sheet (one per participant).
- Prepare copies of the Microscope Lab Procedures (one per group).
- Set up the materials for each group:
    - Small clear jars to collect water samples
    - Microorganism Data Sheet (1 per participant)
    - Microscope Lab Procedures (1 per group)
    - Any print material or internet resources to help with microbe identification (see links for suggested resources)
    - Hand lens
    - Microscope (if available)
        - Microscope slides
        - Coverslips for microscope slides
        - Pipettes
        - Alcohol swabs to clean the eyepiece between each participant
- Set up one or more slide cleaning station(s) with the cleaning supplies you have chosen if slides will be cleaned and reused.

Note: If participants are unable to collect water samples as a class, the facilitator may prepare samples for each group, or the facilitator may request that participants bring in water samples from home. Pond water samples can also be ordered through science companies if needed.

Lesson 4 Microbes in Our Water

Opening Questions

- What do you think is in the water in the jar?
- Is your answer the same after inspecting the water with a hand lens?

Follow-up Question

- Have any of you ever gone swimming in a pond, river, lake, ocean, or another place that was not a swimming pool?

Let's Do It!

1. Allow participants to inspect the container(s) of water. Do not tell participants where the sample(s) came from. Have them describe the water sample to you.

2. Ask the first Opening Question.

    - What do you think is in the water in the jar?

3. Encourage participants at this point to use the hand lenses to explore the water through the container. They should be able to observe small microorganisms moving around in the water.

4. Ask the second Opening Question.

    - Is your answer the same after inspecting the water with a hand lens?

5. Ask the Follow-up Question.

    - Have any of you ever gone swimming in a pond, river, lake, ocean, or another place that was not a swimming pool?

6. Discuss how microorganisms are found in many different water sources.

7. Show the presentation to help participants understand the role microorganisms play in a water source.

8. Explain to participants they will be collecting water samples to observe microorganisms.

Steps to collecting samples:

Part 1 (If applicable-collection of water as a class)

1. Instruct each group to bring at least one copy of the Microbe Data Sheet to record data from the collection location.

2. Give specific instructions for the location from where each group will collect samples.

3. Once at the desired location, direct participants to take the collection container and dip it into the water source and fill it up as much as possible. It is okay if some dirt/debris gets into the sample. Participants are to try to get the water sample from an area where they think there will be a variety of microorganisms.

4. Once the samples have been collected, participants will place the lids on the containers.

5. On the Microbe Data Sheet for each group, participants will take some brief notes about the area from where the sample was collected.

6. Return to the classroom/activity area to complete the next part of the activity.

Part 2

1. Direct participants to move into their small groups.

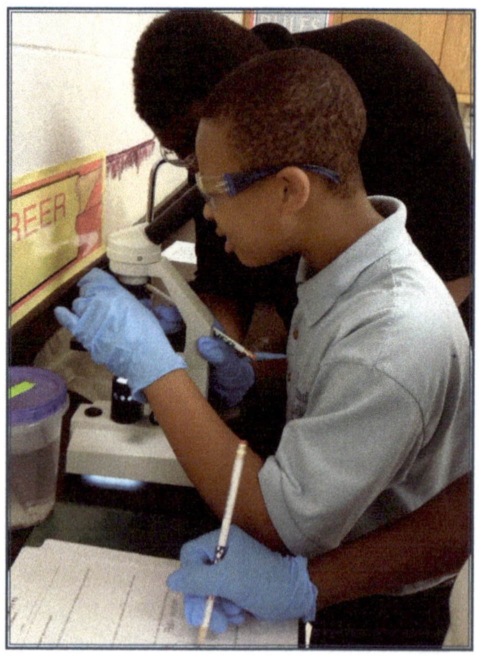

2. Participants will write down some observations about their water sample on their Microbe Data Sheets. They will also include a sketch of what their sample looks like and make predictions about what organisms might be found in the sample.

3. Participants should carefully swirl their water sample to make sure nothing has settled at the bottom.

4. Participants will remove the lid and take a sample from the container using the pipette and then replace the lid on the container.

5. Using one microscope slide, participants will place a drop of the sample into the middle of the slide.

6. Participants will cover the sample with a coverslip or an additional slide.

7. Participants will place the prepared slide under the microscope and adjust until microorganisms are found.

Note: If microscopes are not available, samples can still be taken and placed on a slide, and observed with a hand lens. They can also be observed using a hand lens in the container.

8. Help participants try to identify the microorganisms observed under the microscope using the provided links.

http://www.microscopy-uk.org.uk/pond/

http://science.csumb.edu/~hkibak/305web/documents/pondlife.pdf

http://www.biologycorner.com/worksheets/identifypond.html

http://www.msnucleus.org/watersheds/mission/plankton.pdf

9. Participants will complete the observation pieces (through page 2) about the microorganisms found in the water sample in the appropriate spaces provided on the Microbe Data Sheet.

10. Allow participants to go around the room and observe the water samples collected from other groups. Have them compare to see if they found similar or different microorganisms that were found in the other samples.

11. Direct participants to complete the Comparing Microorganisms page of the Microbe Data Sheet.

12. Conduct a whole-class discussion. Ask participants from different groups to describe and identify the microorganisms found in their samples.

13. Once participants are done with the samples they may be disposed of outside or back into the area from where they were collected. Containers can be washed and reused or discarded.

14. Direct participants to clean their microscopes and slides and complete any other housekeeping duties required at the end of the microscope lab.

Talk It Over

Share 1	**Reflect 2**
What were some of the microorganisms you found in your water sample? Compare the microorganisms you found to the ones found in other groups' samples.	What are some reasons you think other samples had different microorganisms than your group's sample did? Which areas tested do you think have a greater diversity of microorganisms? Why do you think this?
Generalize 3	**Apply 4**
Do you think the number of organisms found in your sample could be found all year round? Why do you think it would change or stay the same?	What would you tell a friend who said it was okay to drink water straight from a pond? Why would you tell them this?

Term and Concept Discovery

Water sample: a sample of water from a specific source

Water testing: testing a sample of water from a specific source for water quality or other factors

Variations

1. Create a poster explaining the different microorganisms that can be found in water.
2. Create a classification chart other participants could use to help identify different microorganisms in water.
3. Use pH strips and test the pH levels in the different water samples. Compare samples to see if the pH level affects the amount and types of organisms present. Alternatively, a red cabbage juice indicator solution may be used to approximate pH measurement. See Appendix B for instructions.
4. Research the microorganisms found. See which microorganisms indicate a healthy water system and which ones represent an unhealthy system. Create a poster showing the differences between the microorganisms.

Lesson 4 Microbes in Our Water

Learn More

These sites can be used to help identify the microorganisms in the water sample:

http://www.microscopy-uk.org.uk/pond/

http://science.csumb.edu/~hkibak/305web/documents/pondlife.pdf

http://www.biologycorner.com/worksheets/identifypond.html

http://www.msnucleus.org/watersheds/mission/plankton.pdf

Education Standard(s)

North Carolina 8th grade Essential Standards:
8.L.1: Understand the hazards caused by agents of diseases that affect living organisms
8.L.2: Understand how biotechnology is used to affect living organisms

National Science Standards:
MS-LS1-1: Conduct an investigation to provide evidence that living things are made of cells; either one cell or many different number and type of cells.
MS-LS1-5: construct a scientific explanation based on evidence for how environmental and genetic factors influence the growth of organisms

Tags: water sampling, microorganisms, microbiology, inquiry learning

Lesson 4 Print Materials

Facilitator Print Materials

None

Group Handouts

Microscope Lab Procedure

Youth Handouts

What's in My Water?!? - Microbe Data sheet

*See page 2 for access to digital assets & extra enrichment.

Microscope Lab Procedure

Materials you will need

- Water sample
- Microorganism Data Sheet (1 per participant)
- Print material or internet resources to help with microbe identification
- Hand lens
- Microscope
- Microscope slides
- Coverslips for microscope slides
- Pipette
- Alcohol swabs to clean the eyepiece between each participant

Making Your Slide

1. Write down some observations about your water sample on the Microbe Data Sheet. Include a sketch of what your sample looks like and make predictions about what organisms might be found in the sample.
2. Carefully swirl your water sample to make sure nothing has settled at the bottom.
3. Remove the lid and take a sample from the container using the pipette. Replace the lid on the container.
4. Using one microscope slide, place a drop of the sample into the middle of the slide.
5. Put a coverslip on top of the sample.
6. Place the prepared slide under the microscope and adjust until microorganisms are found.

Viewing Your Slide

1. Rotate the coarse adjustment knob so that the stage moves down and away from the objective lenses as far as it can go.
2. Rotate the nosepiece of the microscope so that the lowest power objective lens is over the iris diaphragm (this should also be the shortest lens).
3. Place your slide on the stage and center the slide over the hole in the stage.
4. Use the stage clips to secure the slide in place. If you are using a second slide as your coverslip and the stage clips disturb your slide, you may decide not to use the clips for this lab.
5. Using the coarse adjustment knob, move the stage all the way up, as close to the objective lens as it will safely move.
6. Look through the eyepiece. Use the coarse adjustment knob and slowly move the stage down until the prepared slide comes into focus.
7. If you move the stage all the way down and do not ever see anything, move your slide on the stage and repeat steps 5 & 6 again.
8. Once the slide comes into coarse focus, use the fine adjustment knob to bring it into sharper focus.
9. Carefully rotate the nosepiece so that the next higher power objective lens is now pointed at your slide. Do not move either the focus knob or the slide during this step.
10. Move the fine adjustment knob as necessary to bring the slide into sharp focus again.
11. Sanitize the eyepiece of the microscope with an alcohol swab before the next member of your group uses the microscope.
12. Carefully remove your slide from the stage and clean or dispose of it according to the instructions you are given at the appropriate time.

What's in My Water?!? - Microbe Data sheet

Name(s): ______________________________ Date: ______________________________

Collecting your water sample

Describe the area from where your water sample was collected. Include the surroundings plants, animals, etc. that were found in this area.

Looking at your water sample

Describe in detail what your water sample looks like. Include color, contents, and any other observable features.

Sketch what your water sample looks like. Provide details and color if needed:

What microorganisms do you think will be found in this water sample? Why do you think this?

What's in My Water?!? - Microbe Data sheet Continued

Using the microscope or hand lens

Sketch what you saw through the microscope or using the hand lens in the circles below.

Microbe Data Table

Using your references and your observations, record the different microorganisms found in your water sample.

Sketch of Microorganism	Description of Microorganism	What type of microorganism do you think this is?	Interesting fact about this microorganism	Additional notes about microorganism

What's in My Water?!? - Microbe Data sheet Continued

Comparing Microorganisms

1. Compare the microorganisms you found to the ones found in other groups' samples.

2. What are some reasons you think other samples had different microorganisms than your group's sample did?

3. Which areas tested do you think have a greater diversity of microorganisms?
   Why do you think this?

4. Do you think the number of organisms found in your sample could be found all year round?
   Why do you think it would change or stay the same?

5. What would you tell a friend who said it was okay to drink water straight from a pond?
   Why would you tell them this?

Lesson 5 Microbes All Around Us

Introduction

Microbes can be found in many places: on our hands, inside our bodies, and on objects that we use every day. Likewise, they may also be found in the air, water, and soil. Bacteria are the most common living microbes. They are extremely small cells that must be magnified many times to be visible. A microscope is used to view individual bacterial cells. However, colonies of bacteria can grow large enough to be seen without the use of a microscope.

Microbes called fungi are valued for their ability to obtain energy from food sources through a process known as fermentation. This process is vital in the production of alcoholic beverages, bread, and other baked goods because it produces alcohol and carbon dioxide gas.

In some instances, bacteria and fungi can cause various illnesses. If fruits, vegetables, or meats become contaminated by these organisms, they are then unsafe to eat.

Lesson 5 Microbes All Around Us

Key Teaching Points

1. Microbes can be found in many places: on our hands, inside our bodies, and on objects that we use every day.
2. Colonies of bacteria can grow large enough to see without the use of a microscope.
3. Bacteria or fungi can cause various illnesses.

Learner Outcomes

- Compare and contrast microbes as they grow in a nutrient solution.
- Examine various surfaces and graph the bacteria present.
- Create a cartoon/poster showcasing what has been learned from the lesson.

Skill Level: 6-8th grade

Success Indicators

- Participants will create and present a product/deliverable explaining microbial growth.
- Participants will complete a plating activity and graph.

Life Skills

Critical Thinking, Communication, Problem Solving, Teamwork, Presenting

Time Needed

1-5 days

Space

A clear area on table or countertop that has been cleaned thoroughly with a disinfectant

Suggested Group Size

No more than four participants per group

Materials List

- Technology to present the presentation to the class
- Access to technology resources for research

Each individual participant should have:

- 1 subject notebook/journal
- Pen/pencil
- Microbial Math worksheet
- Copies of Participant Data Sheets A, A-1, and A-2
- Microbes All Around Us Lab Reflection (one per participant)
- Parent letter

The items listed below should be made available per group of four:

- One petri dish with agar solution (4-inch size) per participant
- Cotton swab
- Tape
- Disinfectant
- Permanent Marker
- Safety gloves
- Hand lenses
- Alcohol swabs
- Poster board
- Plastic pipette
- Bleach solution

Directions for the disposal of petri dishes:

Once all bacterial growth has been documented on the last day of the experiment, safely destroy the colonies grown. Untape the lid of the plate and then slowly lift the plate at an angle and pipette in about 20-30 mL of a bleach solution. Close and retape the plate. Let the plates stand for at least an hour undisturbed. The plates can then be safely thrown into the trash.

NOTE: *This step MUST be completed with gloves ON.*

Lesson 5 Microbes All Around Us

Before the Activity

- Ensure that you have secured all the necessary materials.
- Prepare technology to show the presentation to the class.
- Prepare copies of the Microbial Math worksheet (one per participant).
- Prepare copies of the Microbes All Around Us Participant Data Sheets A, A-1, and A-2 (one per participant).
- Prepare copies of the Microbes All Around Us Lab Reflection (one per participant).

Note about participant research:
Beginning on Day 2, participants will start a research component for this lesson. Groups will present their research posters on the final day of observation of the petri dishes. Decide ahead of time if the research will be self-directed or if specific areas of research will be assigned to each group.

Areas of focus can include:

- microbial growth
- personal hygiene
- food safety practices
- types of pathogenic microbes
- specific types of illness caused by microbes
- any other topic you feel would be informative for the class as a whole

Participants should link the microbe growth on their petri dishes into the research in some way. This may look different for each group, depending on the topic and what the group has cultured on their petri dishes.

Opening Questions

- Has anyone you know ever had pink eye? Pneumonia? Food poisoning?
- How did they feel when they were infected?
- How did you think they got this disease?
- How did you feel or act around these people when they were infected?

Follow-up questions:

- Does anyone think there are any microbes in this room?
- Where might they be?

Lesson 5 Microbes All Around Us

Let's Do It!

1. Show the presentation to the class.

2. Ask participants if they know why we cook food or keep it in a refrigerator. Explain that putting food in the refrigerator only slows or stops microbial growth; it doesn't kill the microbes. The microbes will grow again when they come out of the refrigerator. The only way to kill microbes is to cook food until it is really hot and cooked all the way through. This is because very warm temperatures kill many harmful microbes.

3. Ask participants about safe handling procedures for food. How should food be handled to maintain safety? Have participants list some safe and unsafe procedures for handling food that is ready to be eaten.

4. Ask the Opening Questions.

    - Has anyone you know ever had pink eye? Pneumonia? Food poisoning?
    - How did they feel when they were infected?
    - How did you think they got this disease?
    - How did you feel or act around these people when they were infected?

5. Facilitate discussion among the participants. Ask participants if they believe the "five-second rule" to be truth or fiction. Allow participants to defend their position.

6. Introduce the Microbial Math worksheet. Follow the scenario on the worksheet to help participants understand the connection between bacterial contamination and growth.

7. Explain that in this lesson, participants will swab various locations that are encountered daily to investigate whether or not diverse microbes are truly present. If present, different types of microbes will be collected from the swabbed samples and transferred to a pre-prepared petri dish. Colonies should be observed growing on the agar over a period of 3-7 days.

8. Ask participants the Follow-up questions.

    - Does anyone think there are any microbes in this room?
    - Where might they be?

9. List their ideas on the whiteboard or chalkboard.

10. Share that the class is going to find out if any microbes are present in the room or even on some of the items that the participants use every day.

11. Divide the participants into four groups of no more than five participants per group. Assign participants in each group one of the following items to swab. Each member of each group will swab and culture a petri dish of their own.

1	Cell Phone or other personal electronic device
2	Belly Button
3	Door Handle
4	Pen, Pencil, or Book
5	Computer Keyboard

Review the Safety Guidelines with ALL participants

1. All materials (book bags, clothing) other than laboratory materials should be off of the work area. NO EATING!
2. Wear the safety gloves provided.
3. Long hair must be tied back.
4. Clean off the workstation with the disinfectant solution (commercial product or 10% bleach solution) BEFORE and AFTER lab activities.
5. Wash hands BEFORE and AFTER lab activities with hand soap.
6. Avoid placing any lab materials in your mouth.

****CAUTION: After the completion of the activity, make sure all participants follow proper lab cleaning protocol. Do not allow the bleach to touch skin, eyes, or clothes. It may burn!**

Day One: Plate Preparation and Swabbing

1. Select one unused petri dish.
2. Label the bottom of the petri dish with the assigned sample being tested and your name.

3. Use a permanent marker to draw an X on the top of the petri dish. This will divide your petri dish into quadrants and make it easier to count any colonies that grow.

4. You will now collect some microbes on the end of a cotton swab.

5. Roll the swab across your designated item or area and then lightly draw a wavy line with it on the gelled agar.

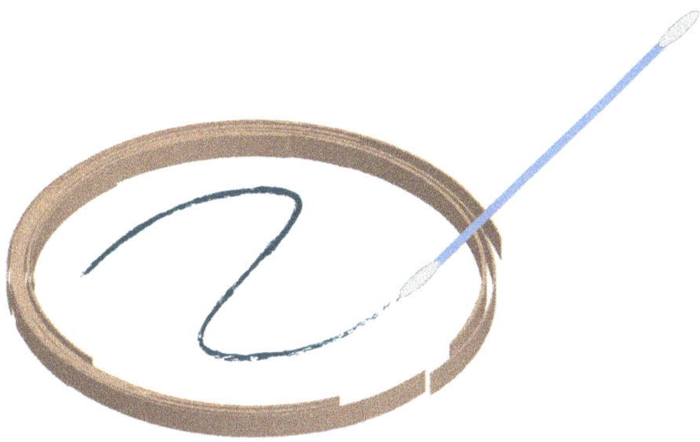

6. Place the top on the petri dish. For your protection, tape the petri dish closed. For safety reasons, **DO NOT** re-open the petri dish; you could be culturing a pathogen. You can view the growing microbes through the clear dish.

7. Turn the plate upside down and then place the plate in a warm, dark place to grow for three days. You can look forward to seeing a variety of microbes growing in a few days. As time increases, more colonies will grow and increase in size.

NOTE: Any temperature up to about 98° degrees F (37° degrees C) should be fine. Examples of warm places include, but are not limited to: the top of a refrigeration unit, a sunny window, or the inside of a drawer. If this activity is being conducted during the winter months, consider setting up an incubation area with a grow lamp or another warming device that can be monitored for safety while you are present. Warming devices may be turned off when unattended.

Day Two-Day Four: Research and Documentation and Observation of Growth

1. Begin the research portion of this lesson. In their small groups, participants will conduct research as assigned. Suggestions for participant research are listed in the Before the Activity section above.

2. Explain to participants they are to display the results of their research as a group. Each display should include at least a one page

summary of the group's findings. Encourage participants to organize their displays as creatively as possible. They can include comic strips, infograms, diagrams, charts, graphs, or any other method that conveys the information in an interesting and informative way. Alternatively, participants may create an electronic presentation as technology is available.

3. Examine the petri dishes each day at the same time WITHOUT opening them.

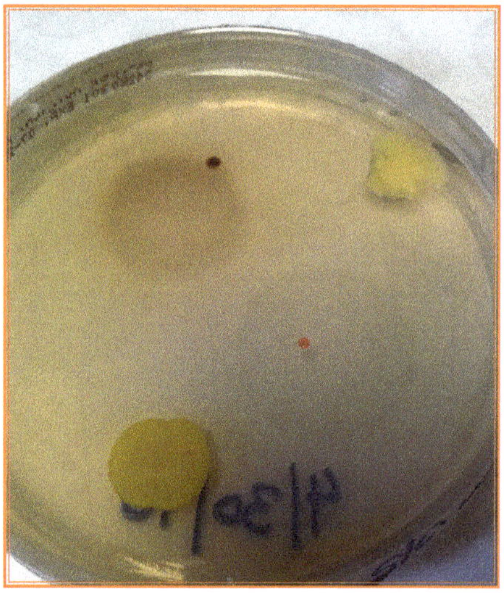

4. Record all data on Participant Data Sheet A each day.

5. Compare and discuss the growth results with two other participants each day. Document their results on Participant Data Sheets A-1 and A-2.

6. Compare your results to the findings at:

   http://robdunnlab.com/projects/belly-button-biodiversity/

Day Five (or end of observations): Present and Network Like A Scientist

1. Ask the Share questions.

   - What has changed inside the petri dishes?
   - Were you surprised about the microbes that grew in the area your group tested? Why or why not?
   - Why do you think your area had a significant growth of microbes?
   - Comparing your data to the other groups, which area tested seemed to have the most microbial growth? Why do you think this occurred?

- Which area had the least microbial growth? Why do you think this occurred?

2. Ask the Generalize questions.

   - Do your results prove or disprove that there are microbes everywhere?
   - With all these bacteria growing, why do you think people aren't more sick or sick more often?
   - What does the information gained from this experiment tell you about human hygiene?
   - What new questions did your results bring up?

3. Give each group a few minutes to present their posters as well as to observe those of their peers.

4. Discuss and document the similarities and differences among their findings.

5. Ask the Apply questions.

   - What are some measures people could take to prevent the spread and growth of microbes?
   - From this activity and your research, will you change any of your hygiene practices? If so, which one(s) and why? Explain your answers.
   - Have you noticed signs in almost all public restrooms stating that all employees must wash their hands before returning to work? Why is this important and how is it relevant to what you have learned in from this lesson?

6. Participants will complete the Lab Reflection.

Lesson 5 Microbes All Around Us

Talk It Over

Share 1

What has changed inside the petri dishes?

Were you surprised about the microbes that grew in the area your group tested? Why or why not?

Why do you think your area had a significant growth of microbes?

Comparing your data to the other groups, which area tested seemed to have the most microbial growth? Why do you think this occurred?

Which area had the least microbial growth? Why do you think this occurred?

Reflect 2

Complete the Microbes All Around Us Lab Reflection.

Microbes All Around Us Lab Reflection

Directions: Fill in the blanks with your answers.

Observation:

I claim that when

then

I know this because I observed

Conclusion:

Today, I learned

Originally I thought

Reflection:

What really surprised me about this investigation was

Extensions / Further Investigation(s):

After doing this investigation, I now wonder

We found

but I still want to know

Generalize 3

Do your results prove or disprove that there are microbes everywhere?

With all these bacteria growing, why do you think people aren't more sick or sick more often?

What does the information gained from this experiment tell you about human hygiene?

What new questions did your results bring up?

Apply 4

What are some measures people could take to prevent the spread and growth of microbes?

From this activity and your research, will you change any of your hygiene practices? If so, which one(s) and why? Explain your answers.

Have you noticed signs in almost all public restrooms stating that all employees must wash their hands before returning to work? Why is this important and how is it relevant to what you have learned in from this lesson?

Variations

1. Slices of cooked potatoes can be used as an alternative to nutrient agar. Boil whole potatoes until almost soft. Using a clean, dry knife, cut potatoes into 1/4-in. slices. Place each slice in a Petri dish or a clean, resealable, plastic bag. To safely discard potatoes in the bags after the activity, pour about 20 mL of a 10% bleach solution into each bag. Seal and discard the bags.

2. Have participants design additional experiments to test for the presence of microbes. They might examine water from different sources, compare washed and unwashed hands, or see which kinds of food grow the most kinds of microbes or spoil most quickly.

3. Have participants investigate what happens when similar samples are grown at room temperature and in the refrigerator. Based on their results, conduct a discussion about the importance of refrigerating leftover food.

4. Participants can record their findings in a table on the board or chart paper to compare class data. Results can be used to determine which place seemed to have the most bacterial colonies appear.

5. Clean/sterilize the areas tested, and then try the experiment again to see the effectiveness of the sanitization and how fast microbes can reproduce.

6. Instead of counting the colonies, participants can measure the size of the colonies in centimeters and graph the growth on a graph.

7. Have participants take an online assessment (they may use their mobile device or computer with internet access) by visiting this website: http://theoatmeal.com/quiz/phone_germs.

Acknowledgments

http://robdunnlab.com/projects/belly-button-biodiversity/

Lesson 5 Microbes All Around Us

Term and Concept Discovery

Bacilli: any rod-shaped or cylindrical bacterium of the genus Bacillus, comprising spore-producing bacteria.

Cocci: a spherical bacterium

Eukaryote: any organism having as its fundamental structural unit a cell type that contains specialized organelles in the cytoplasm, a membrane-bound nucleus enclosing genetic material organized into chromosomes, and an elaborate system of division by mitosis or meiosis, characteristic of all life forms except bacteria, blue-green algae, and other primitive microorganisms

Hygiene: a condition or practice conducive to the preservation of health, as cleanliness.

Organism: a form of life composed of mutually interdependent parts that maintain various vital processes

Prokaryote: any cellular organism that has no nuclear membrane, no organelles in the cytoplasm except ribosomes, and has its genetic material in the form of single continuous strands forming coils or loops, characteristic of all organisms in the kingdom Monera, as the bacteria and blue-green algae

Protists: any of various one-celled organisms, classified in the kingdom Protista, that are either free-living or aggregated into simple colonies and that have diverse reproductive and nutritional modes,including the protozoans, eukaryotic algae, and slime molds: some classification schemes also include the fungi and the more primitive bacteria and blue-green algae or may distribute the organisms between the kingdoms Plantae and Animalia according to dominant characteristics.

Spirilli: any of several spirally twisted, aerobic bacteria of the genus Spirillum, certain species of which are pathogenic for humans

Virus: a microbe that replicates itself by taking over cells in a host organism. At this time, there is debate among scientists as to whether viruses are actually considered to be living organisms. Unlike pathogenic bacteria, viruses cannot be killed using antibiotic medicines.

Learn More

Additional references to the belly button project

http://www.yourwildlife.org/

http://www.npr.org/blogs/krulwich/2011/07/09/137653646/venus-has-a-crowded-belly-button

http://www.microbiologyonline.org.uk/about-microbiology/introducing-microbes/bacteria

Content related articles/videos/games

http://newsfeed.time.com/2012/10/27/study-your-cellphone-is-teeming-with-germs/

http://microbemagic.ucc.ie/about_microbes/microbes.html

http://tlc.howstuffworks.com/home/how-to-clean-your-cell-phone.htm

Lesson 5 Microbes All Around Us

Education Standard(s)

North Carolina 8th grade Essential Standards:
8.L.1: Understand the hazards caused by agents of diseases that affect living organisms
8.L.2: Understand how biotechnology is used to affect living organisms

National Science Standards:
MS-LS1-1: Conduct an investigation to provide evidence that living things are made of cells; either one cell or many different number and type of cells.
MS-LS1-5: construct a scientific explanation based on evidence for how environmental and genetic factors influence the growth of organisms

Math Common Core Standards:
Objective 1.06: Use mathematics to gather, organize, and present quantitative data resulting from scientific investigations:

- Measurement.
- Analysis of data.
- Graphing.
- Prediction models.

Tags: microbiology, science, bacteria, plating, biotechnology, graphing

Lesson 5 Print Materials

Facilitator Print Materials

None

Group Handouts

None

Youth Handouts

Microbial Math worksheet

Microbes All Around Us Participant Data Sheet A, A-1, A-2

Microbes All Around Us Lab Reflection

*See page 2 for access to digital assets & extra enrichment.

Microbial Math worksheet

Bob dropped two sticky, gooey donuts on the floor of his kitchen. Quickly bending to pick them up, he replaced one in the box and picked the less smushed of the two to take along with him as he dashed to catch his ride for school at 7:30 am.

Bob was, of course, eating not only the donut but also many bacteria and other particles the donut picked up from the kitchen floor. Luckily for Bob, his immune system was strong, the kitchen floor had been mopped recently, and the pathogens he ingested did not end up causing him any illness.

But what about the donut he put back in the box? What threat does that donut pose to his family?

While on the floor, the donut picked up five *Staphylococcus* bacteria. The bacteria stayed on the donut and settled into the box, where they decided to become comfortable with their new environment (the donut). For approximately ten hours, the bacteria didn't grow or move - then the explosion began. For the next several hours, until Bob needed an evening snack, the Staph bacteria doubled every thirty minutes.

Beginning at 5:30 pm the bacteria were doubling every thirty minutes, and by 8:00 pm Bob was hungry for an evening snack. How many bacteria did Bob eat? Complete the following math problem to find out.

6:00pm ________ bacteria × 2 = ________

6:30pm ________ bacteria × 2 = ________

7:00pm ________ bacteria × 2 = ________

7:30pm ________ bacteria × 2 = ________

8:00pm ________ bacteria × 2 = ________

Graph the growth of the bacteria on the donut from the time Bob dropped it until he ate it. :)

Microbial Math worksheet Continued

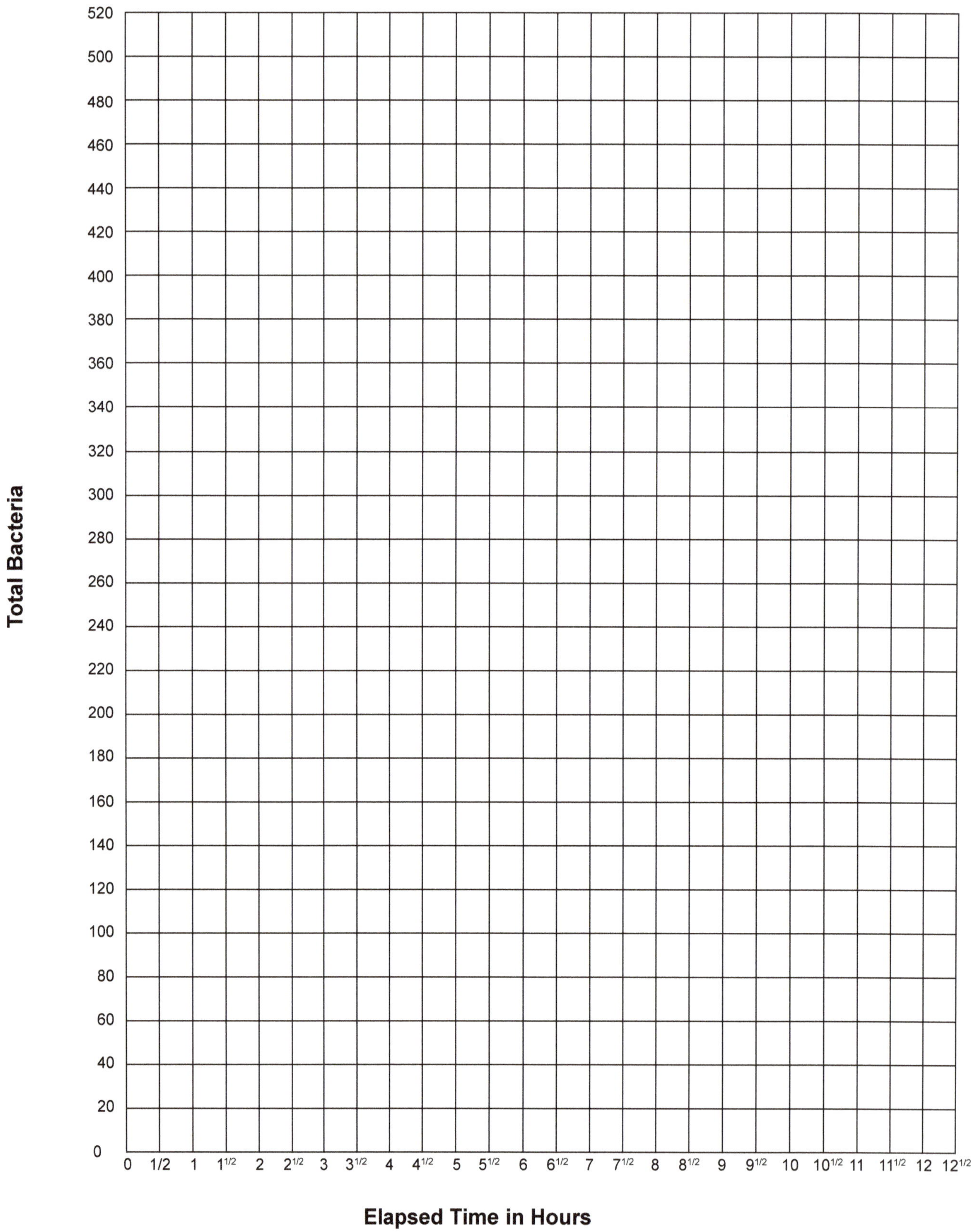

Microbes All Around Us
Participant Data Sheet A: Documenting Microbe Growth

Purpose: ____

____

Question: ____

____

Hypothesis: ____

____

____

Complete the data table below for your swabbed ____________ (list the sample) sample observed.

	What size are **your** colonies (cm)?	How many colonies are present?	What colors are **your** colonies?	What shapes are **your** colonies?	Any other observations? You may want to include a drawing of any growth on your plate.
Day 1					
Day					
Day					
Day					

Microbes All Around Us
Participant Data Sheet A-1: Documenting Microbe Growth

Examine the growth on another person's plate.

First Partner's Name: ______________________ Sample Observed: ______________

	What size are **these** colonies (cm)?	How many colonies are present?	What colors are **these** colonies?	What shapes are **these** colonies?	Any other observations? You may want to include a drawing of any growth on this plate.
Day 1					
Day					
Day					
Day					

Explain the results: Were there any similarities? Differences?

____

____

____

____

____

Microbes All Around Us
Participant Data Sheet A-2: Documenting Microbe Growth

Examine the growth on another person's plate.

First Partner's Name: ______________________ Sample Observed: ______________

	What size are **these** colonies (cm)?	How many colonies are present?	What colors are **these** colonies?	What shapes are **these** colonies?	Any other observations? You may want to include a drawing of any growth on this plate.
Day 1					
Day					
Day					
Day					

Explain the results: Were there any similarities? Differences?

____

____

____

____

____

Microbes All Around Us Lab Reflection

Directions: Fill in the blanks with your answers.

Observation:

I claim that when ______________________________,

then ______________________________.

I know this because I observed ______________________________.

Conclusion:

Today, I learned ______________________________

______________________________.

Originally I thought ______________________________

______________________________.

Reflection:

What really surprised me about this investigation was ______________________________

______________________________.

Extensions / Further Investigation(s):

After doing this investigation, I now wonder ______________________________

______________________________.

We found ______________________________,

but I still want to know ______________________________.

Lesson 6 DNA Extraction Activity

Introduction

All microbes have deoxyribonucleic acid (DNA) and so do you! DNA is the genetic information that guides the development and functioning of all living things. Many aspects of genetic research require the extraction and isolation of DNA from cells of various organisms.

DNA extraction is the process scientists follow to remove the genetic information from organisms to gain an understanding of their genetic properties. DNA extraction is used for many reasons in the microbiology field. Scientists can look at the DNA of an organism to understand its genetic information as well as its common ancestors.

In this activity, participants will gain a full understanding of how DNA can be extracted from various fruits and vegetables. Participants will use common household products to remove the DNA strands from these items and compare the different DNA strands. This will allow participants a chance to observe the genetic information and understand how this relates to the microbiology profession.

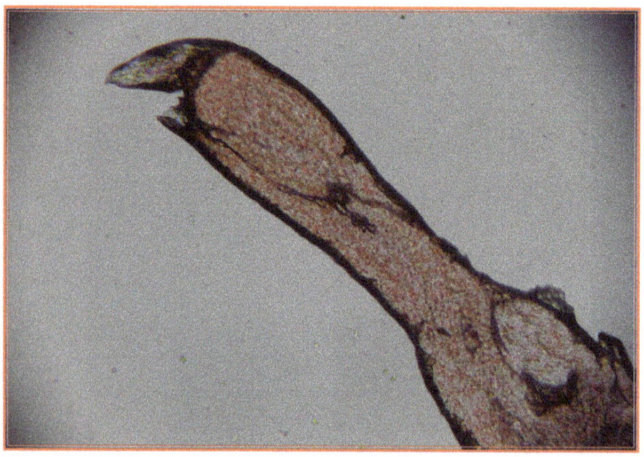

Tomato DNA

Kiwi DNA

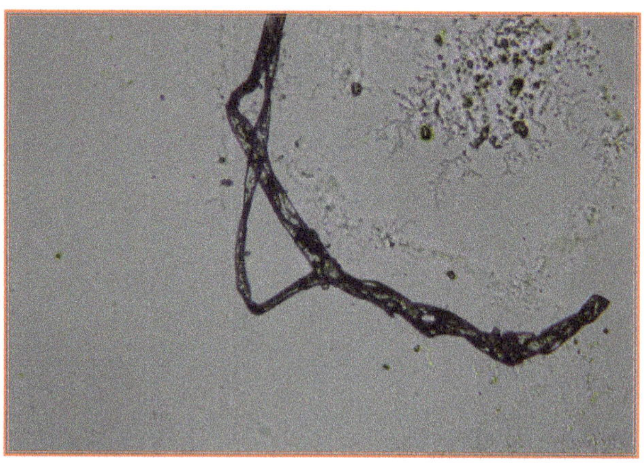

Strawberry DNA

Lesson 6 DNA Extraction Activity

Key Teaching Points

1. All living organisms have deoxyribonucleic acid (DNA), including microbes.
2. DNA is the genetic information that guides the development and functioning of all living things.
3. DNA extraction is the process scientists follow to remove the genetic information from organisms to gain an understanding of their genetic properties.

Learner Outcomes

- Experience how to extract DNA from different fruits and vegetables and compare their findings.
- Gain an understanding of the importance of DNA extraction and how microbiologists use DNA in their careers.

Skill Level: 6-8th grade

Success Indicators

- Participants will understand the process of DNA extraction in its relation to microbiology.
- Participants will successfully extract DNA from a sample of fruit or vegetable.

Life Skills

Teamwork, Critical Thinking, Communication, Problem Solving

Time Needed

1 Day (90 minutes)

Space

Classroom or area where participants can work in groups

Suggested Group Size

4-5 participants per group

Materials List

For Facilitator:

- Technology to present the presentation to the class
- Copy of DNA Extraction Activity Participant Lab Sheet
- Materials listed below for each group/participant
- Knife to peel any fruit as necessary

For each participant in group:

- Copy of DNA Extraction Activity Participant Lab Sheet
- Copy of Observation and Reflections handout
- Coffee Filter
- Bathroom cup (5 oz)
- 1 cup of warm water (tap water)
- Piece of the selected item being tested
- Plastic Spoon
- Toothpick
- Quart sized sealable plastic bag
- Small test tube with lid
- Plastic Pipette
- Hand lens

Materials needed for each group:

- Bottle of clear liquid dish detergent
- Container of salt
- 8oz cup full of ice and water
- Extra plastic bags in case any group members break their bags during the experiment
- Bottle of 91% rubbing alcohol (kept in freezer)
- Colored pencils
- Funnel
- Measuring spoon set
- Test tube brushes

Keep the alcohol in the freezer for at least 2 hours until just before this lesson begins.

Lesson 6 DNA Extraction Activity

Before the Activity

- Make sure the materials listed above are ready and available for the participants to complete the DNA extraction activity.
  Note: Frozen fruit will work for this learning experience.
- Prepare technology to show the presentation to the class.
- Prepare copies of the DNA Extraction Activity Participant Lab Sheet for each participant.
- Prepare copies of the Observation and Reflection handout for each participant.
- Prepare one copy of the DNA Extraction Activity Participant Lab Sheet for facilitator use.
- Peel any fruit or vegetable that has thick skin before distributing it to the participants.
- Place the alcohol in the freezer at least two hours before use.
- Print or prepare a model of DNA to show participants.

Opening Questions

- What is this a picture/model of?
- Why do you think if it is inside of every tiny cell in your body that we can know what it looks like?

Follow-up questions:

- What are some organisms that you think have very similar DNA?
- If they have similar DNA, what does this mean about some of their ancestors?
- How do you think scientists extract DNA from organisms?
- Do you think this is a difficult process? Why or why not?
- Why do you think it is important for us to know how to extract DNA from various organisms?

Let's Do It!

1. Bring in a model or a picture of DNA on the day of the activity. Tell participants that this model/picture is inside every single cell in their body.

2. Ask the Opening Questions.

    - What is this a picture/model of?
    - Why do you think if it is inside of every tiny cell in your body that we can know what it looks like?

3. Continue to explain to participants that many organisms have similar DNA to other organisms and it only takes a slight change in the DNA sequence to make a completely new organism.

4. Ask the first two Follow-up questions.

- What are some organisms that you think have very similar DNA?
- If they have similar DNA what does this mean about some of their ancestors?

5. Explain to participants that scientists have figured out ways to extract DNA from organisms so they can compare their genetic information to other living things. Tell participants that this is how we have been able to learn a great deal about organisms that lived many years ago and also find modern-day organisms that they are closely related to.

6. Ask the final Follow-Up Questions.

    - How do you think scientists extract DNA from organisms?
    - Do you think this is a difficult process? Why or why not?
    - Why do you think it is important for us to know how to extract DNA from various organisms?

7. Explain to participants that today they are going to experience how to extract DNA from different fruits and vegetables and compare their findings to others. Participants will also gain a better understanding of the importance of DNA extraction and how microbiologists use DNA in their careers.

8. Show the presentation to the class. Have participants engage in discussion about the information and the purpose of DNA extraction.

9. Divide the participants into groups of 4 or 5 and assign them a group number and an item to test as indicated in the chart below:

10. Direct participants to form their groups.

1	Kiwi
2	Strawberry
3	Crushed Tomatoes
4	Bananas
5	Peaches

11. Read over the instructions for the DNA extraction with the class as necessary.

12. Once the groups have finished their experiments, have participants complete the Observation and Reflection handout questions.

13. As applicable, ask participants to share and discuss their answers from their Observation and Reflection handout with the class.

14. Recycle all test tubes used in the experiment for use again by washing using warm water, detergent, and test tube brushes. Rinse each test tube thoroughly with warm water followed by alcohol. Turn the test tubes upside down and allow them to dry on a paper towel.

15. Clean up the work area and return materials to the proper locations.

Talk It Over

Share 1	Reflect 2
Compare the DNA strand from your fruit/vegetable to another participant's DNA strand who also used the same item. How does your strand differ or look the same as the other person's strand? Compare the DNA strand from your fruit/vegetable to another participant's DNA strand who tested a completely different item. Share the similarities and differences between the two DNA strands.	Why do you think some groups extracted more DNA out of their item than others? Explain. What do you think the purpose of smashing up the fruit/vegetable was? Do you think this made it easier to extract the DNA? Explain. How would our lives be different without the advances made with microbiology and DNA extraction?
Generalize 3	**Apply 4**
What general statements could you make about how DNA is extracted from organisms and why DNA extraction is important in the biotechnology field?	How would you explain what DNA is and the importance of DNA to a friend? How would you explain the reason DNA extraction and testing is done on organisms?

Variations

1. Instead of completing step four (4) of the Lab Sheet (smashing/ grinding fruits of vegetables), once peeled, place the object in a blender on the high-speed setting for about 20 seconds. Then, transfer the contents into a sealable plastic bag and proceed with the rest of the protocol.

2. Remove some of the DNA and place it on a micro slide. Then, stain it with methylene blue and add one drop of water to it. Cover it with a coverslip. Next, view the DNA slide under a microscope.

3. You may want to try fruits and vegetables other than the ones listed, including but not limited to: split peas, seedless water-melon, or chickpeas. You may also want to consider contacting your local or state farmer's market to see which fruits/vegetables are in season.

4. Have participants research and create models of the DNA molecule (structure) using pipette cleaners, origami templates, playdough, coloring pages, or other methods that can be found via the web.

References:

https://youtu.be/oDMi4SWMq0c

Term and Concept Discovery

Biotechnology: the use of living organisms or other biological systems in the manufacture of drugs or other products or for environmental management, as in waste recycling: includes the use of bioreactors in manufacturing, microorganisms to degrade oil slicks or organic waste, genetically engineered bacteria to produce human hormones, and monoclonal antibodies to identify antigens.

DNA: Genetics. deoxyribonucleic acid: an extremely long macromolecule that is the main component of chromosomes and is the material that transfers genetic characteristics in all life forms,constructed of two nucleotide strands coiled around each other in a ladderlike arrangement with the side pieces composed of alternating phosphate and deoxyribose units and the rungs composed of the purine and pyrimidine bases adenine, guanine, cytosine, and thymine: the genetic information of DNA is encoded in the sequence of the bases and is transcribed as the strands unwind and replicate.

Lesson 6 DNA Extraction Activity

Learn More

Content related articles
https://sites.ced.ncsu.edu/stem-career-awareness/

Education Standard(s)

North Carolina 8th grade Essential Standards:
8.L.2: Understand how biotechnology is used to affect living organisms

National Science Standards:
MS-LS1-1: Conduct an investigation to provide evidence that living things are made of cells; either one cell or many different number and types of cells
MS-LS1-5: Construct a scientific explanation based on evidence for how environmental and genetic factors influence the growth of organisms

Tags: microbiology, science, bacteria, plating, biotechnology, graphing

Lesson 6 Print Materials

Facilitator Print Materials

Copy of Participant Handouts for facilitator use

Group Handouts

None

Youth Handouts

DNA Extraction Activity Participant Lab Sheet
Observation and Reflection handout

*See page 2 for access to digital assets & extra enrichment.

All microbes have deoxyribonucleic acid (DNA) just like you and I!

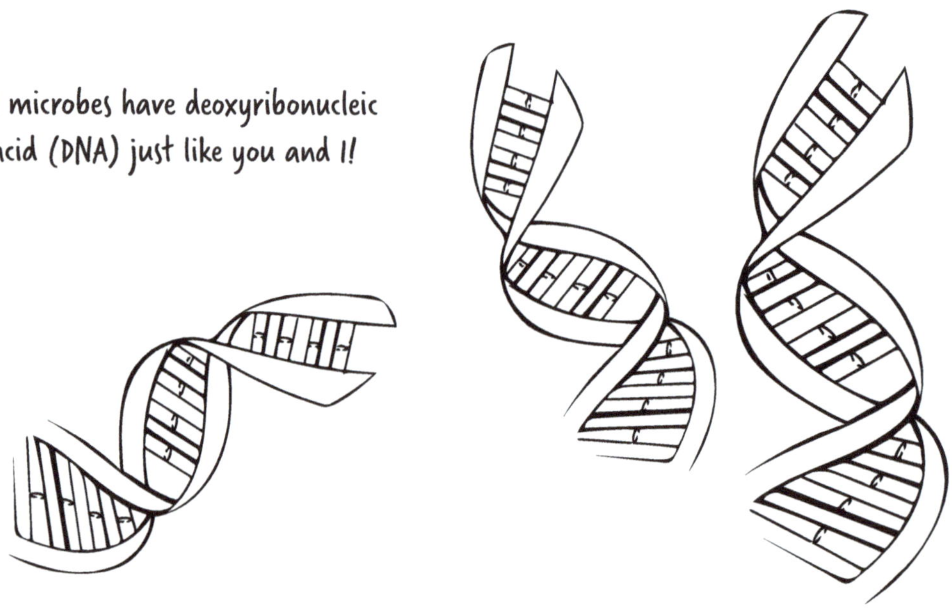

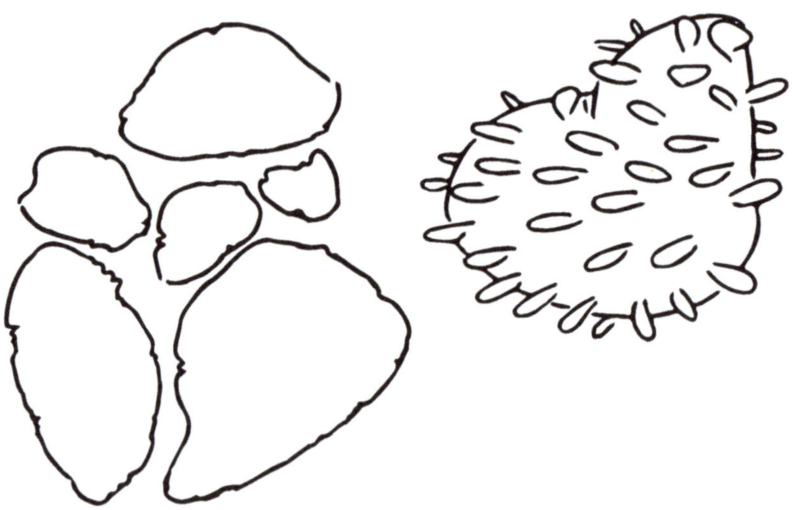

In this last lesson, we are going to learn about the genetics of microbes and how they differ.

DNA Extraction Activity Participant Lab sheet

Name: ______________________________ Date: ______________

Item Being Tested: ______________________________

Introduction

Microbes have deoxyribonucleic acid (DNA) and so do you! DNA is the genetic information that guides the development and functioning of all living things. Many aspects of genetic research require the extraction and isolation of DNA from cells of various organisms. In this lab, you will use household materials to extract DNA from a piece of fruit or vegetable.

Materials

For each participant in group:

Copy of Lab sheet	Plastic spoon
Coffee filter	Toothpick
Bathroom cup (5oz)	Quart sized sealable plastic bag
1 cup of warm water	Small test tubes with lids
Piece of the selected item being tested	Plastic pipette

Materials

For whole group:

Bottle of clear liquid dish detergent	Bottle of 91% rubbing alcohol (kept in the freezer)
Container of salt	Funnel
8 oz cup full of ice and water	Measuring spoon set
Hand lens	

Procedure

1. Acquire the selected fruit or vegetable from your instructor.
2. Make sure your group has the materials listed above.
3. Break your fruit or vegetable into pieces and place it in the plastic zip bag.
4. CAREFULLY grind up the item using your hands and fists until the mixture is free of any lumps. It should have the consistency of baby food.

DNA Extraction Activity Participant Lab sheet Continued

**Make sure you are being careful not to break the bag during this process. The purpose of mashing the fruit/vegetable is to rupture the cell membranes and to release the molecules contained within the cell and within its organelles.

5. Measure 2 tablespoons of detergent into the bag.
6. Measure ½ teaspoon of salt into the bag.
7. Close the bag carefully.
8. Mix the detergent and salt into the fruit and vegetable mixture to make sure it is thoroughly spread throughout the mixture.
9. Measure ½ cup of warm water into the plastic zip bag. Close the bag carefully again. Mix the water thoroughly with the other contents.
10. Allow the mixture to sit for 5-10 minutes to allow the contents to fully combine.
11. Place the coffee filter inside of the funnel and place the funnel over the top of the bathroom cup.
12. Carefully pour the mixture from the bag into the funnel. The coffee filter will allow only the liquid portion to go into the cup.
13. Using your fingers, carefully mash the mixture caught by the filter to make sure all the liquid drips into the cup.
14. Remove the funnel and dispose of the coffee filter.
15. Carefully place the filtered solution into the test tube until it is 1/3 full.
16. Place the cap on the test tube and cool the mixture in the water/ice bath for 5 minutes.
17. Remove the test tube from the ice bath. Carefully remove the top.
18. Using the pipette, slowly add the COLD alcohol to the test tube, filling it ¾ of the way full.
    ***Hold the test tube at an angle when adding the alcohol.
19. Bubbles should start forming and you should see the DNA precipitate (float) out of the solution near the top (it will appear as a clear goopy substance).
20. Observe the DNA using a hand lens.
21. Draw what you observe in the Reflections part of this sheet. Compare your results to other participants' results. Look at similarities and differences between the DNA strands.
22. The DNA may be transferred to a screw-cap test tube and stored at 4O°C for future use in later experiments. If keeping the DNA as a souvenir, place it in a transparent, sealable container with 70-91% ethyl alcohol.
23. Clean up your work area and return all materials to their proper locations and complete the Reflections/Follow-up section of your lab sheet.

NOTES:

Observation and Reflection

Name: ______________________________ Date: ______________

1. In the space below draw what the contents of your baggie looked like before and after you smashed up your item.

BEFORE

AFTER

2. What do you think the purpose of smashing up the fruit/vegetable was? Do you think this made it easier to extract the DNA? Explain.

3. In the space to the side draw what you observed in your test tube once the alcohol was added.

Observation and Reflection Continued

4. Draw in the space provided what your DNA strand looked like through the hand lens.

5. Were you successful in extracting the DNA from your item? How do you know?

6. Compare the DNA strand from your fruit/vegetable to another participant's DNA strand who also used the same item. How does your strand differ or look the same as the other person's strand?

7. Compare the DNA strand from your fruit/vegetable to another participant's DNA strand who tested a completely different item. Share the similarities and differences between the two DNA strands.

8. Why do you think some groups extracted more DNA out of their item than others? Explain.

Observation and Reflection Continued

9. How would our lives be different without the advances made with microbiology and DNA extraction?

10. What general statements could you make about how DNA is extracted from organisms and why DNA extraction is important in the biotechnology field?

11. How would you explain what DNA is and the importance of DNA to a friend?

12. How would you explain the reason DNA extraction and testing is done on organisms?

NOTES:

Appendix A

Grab and Go Lessons

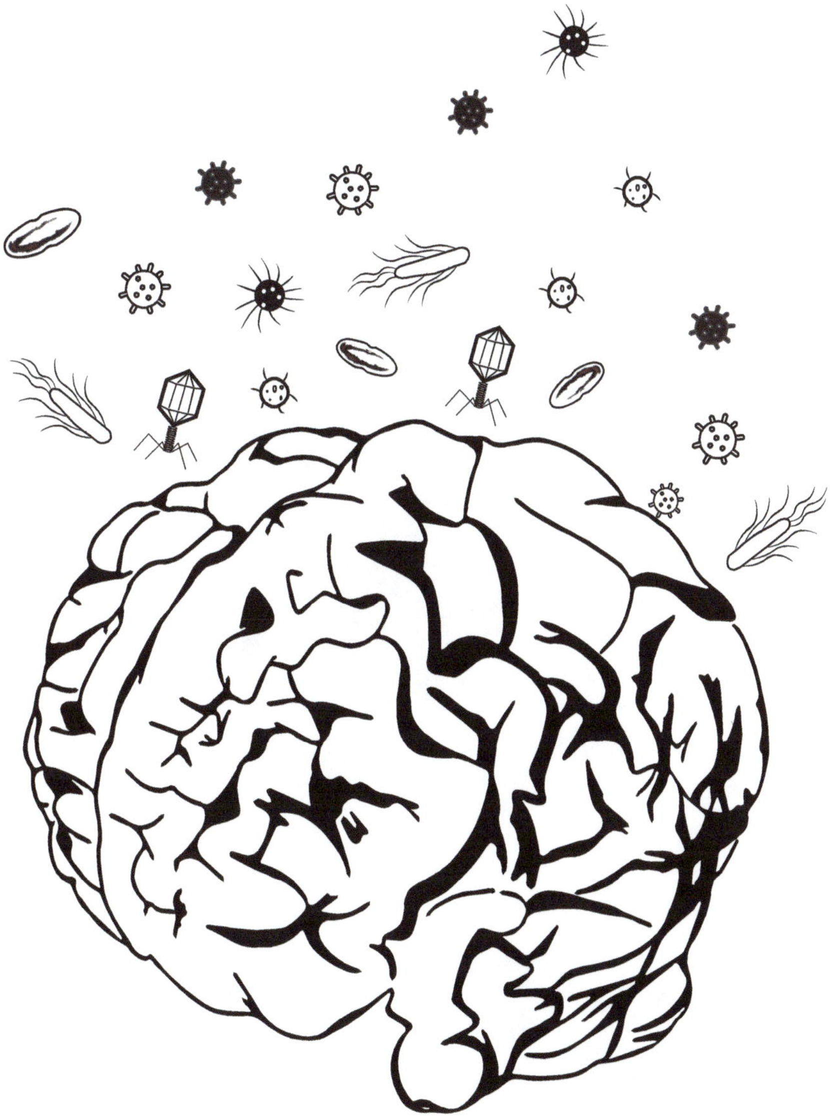

Grab and Go 1 Introducing the Microscope

Introduction

In this quick lesson, participants will learn the parts and functions of the microscope. Two activities in the middle school microbiology curriculum require participants to be able to create and view slides. By completing this activity, participants should be well prepared to tackle those lessons.

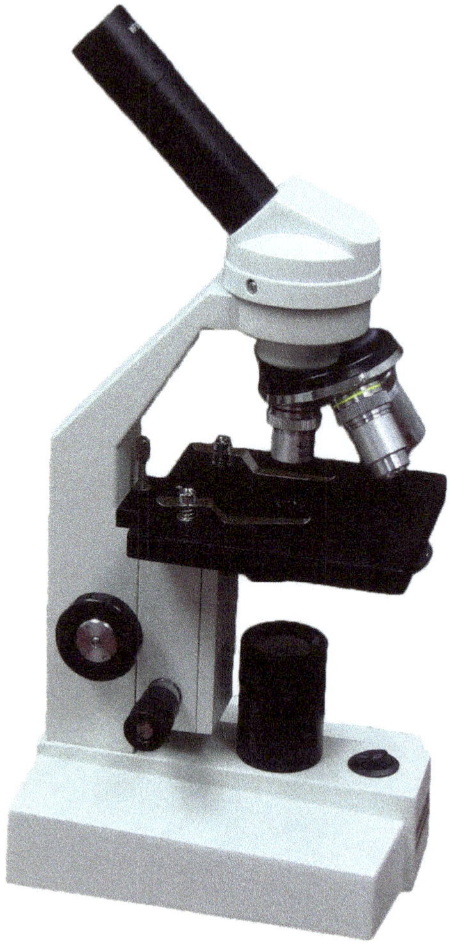

APPLE SLICE

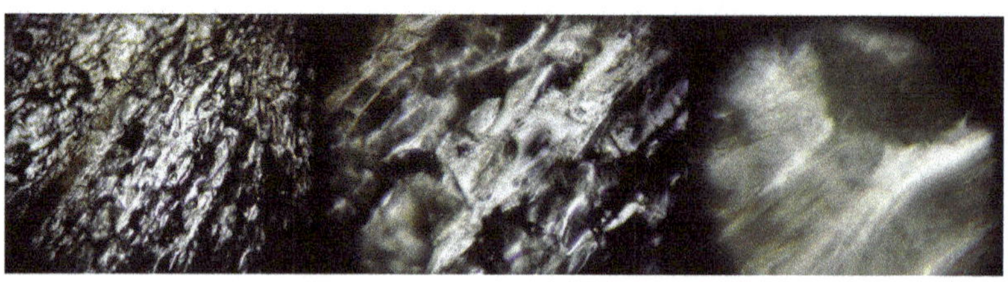

40x → 100x → 400x

Grab and Go 1 Introducing the Microscope

Key Teaching Points

1. Microscopes magnify visible or invisible objects large enough to view and study.
2. Microscopes are made of specific parts with specific functions.
3. Specific steps are required to prepare a slide for microscopy.

Learner Outcomes

- Identify and label the parts of a microscope.
- Prepare a slide for microscopy.
- Properly operate the parts of a microscope to view an image on a slide in focus.

Skill Level: 6-8th grade

Success Indicators

- The participants will be able to properly use a microscope in a classroom lab setting.

Life Skills

Teamwork, Critical Thinking, Communication, Problem Solving

Time Needed

30-45 minutes

Space

Classroom or meeting area able to accommodate small groups of participants using microscopes

Suggested Group Size

3-4 participants per group

Materials List

Group Supplies:

- Microscope (one per group, or one for the class if more are unavailable)
- Magnifying glass
- 4-12 Slides
- Coverslips (or additional slides to act as coverslips)
- Dropper
- Small cup of water
- Paper towels or napkins to use to absorb water
- Alcohol swabs to clean the eyepiece between each participant
- Microscope Practice Lab instructions

Class:

- "Introduction to Your Microscope" video
- Technology to present the instructional video to the class
- Label the Microscope worksheet (one per participant)
- Slide cleaning materials for after the lab if the slides will be reused
    - **Options:** soap and water, glass cleaner, alcohol spray, or lens cleaner AND paper towels, tissues, or lint-free tissues

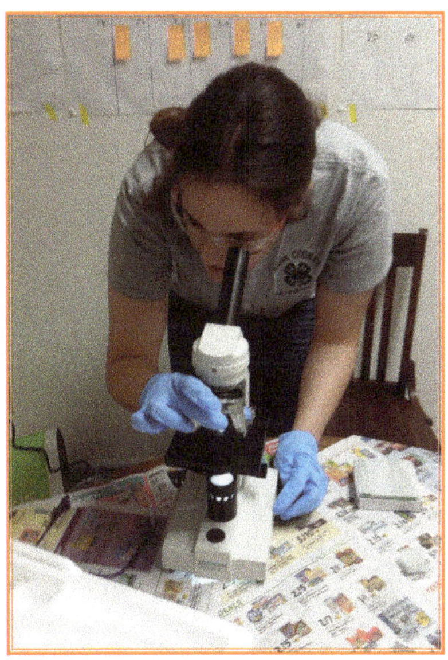

Grab and Go 1 Introducing the Microscope

Before the Activity

- Prepare technology to show the instructional video to the class.
- Prepare copies of the Label the Microscope worksheet for each participant.
- Set up one or more slide cleaning station(s) with the cleaning supplies you have chosen if slides will be cleaned and reused.
- Assign participants to small groups depending on how many microscopes are available. Groups of 3-4 are optimal for this learning experience.
- Arrange microscope stations for each small group.
    - Microscope
    - Magnifying glass
    - 4-12 Slides
    - Coverslips (or additional slides)
    - A small cup of water
    - Dropper
    - Alcohol swabs
    - Microscope Practice Lab worksheet
- Optional: Choose and collect other substances ahead of time that participants may use to create slides, or provide previously prepared slides for participants to view.

Opening Questions

- How do we study objects that are too small for us to see?

Follow-up questions:

- What parts of this microscope are familiar to you?
- Can you identify any of the parts of the microscope?
- Can you explain the function of any of the parts of the microscope?

Let's Do It!

1. Ask the Opening Question. Participants may provide various answers, including microscopes, magnifying glasses, binoculars, or taking a picture and zooming in (digital photography).

2. Tell the participants they will be using microscopes in this learning experience.

3. Ask participants to explain the word "microscope" by breaking the word into two parts: "micro" and "scope." Elicit or tell that "micro" means small, while "scope" means to look or see.

4. Hand out the Label the Microscope worksheet.

5. Ask the Follow-Up Questions.

- What parts of the microscope are familiar to you?
- Can you identify any of the parts of the microscope?
- Can you explain the function of any of the parts of the microscope?

6. Show the Microscope Video.

7. Partway through the video, stop the video and allow participants to transfer the wording on the screen to the Microscope Labels worksheet. [Note: If you are using a microscope that is significantly different from the one in the video, you may choose to substitute a different worksheet or have the participants draw the microscope and its parts.]

8. At the end of the Microscope video, lead the participants in determining the total magnification of the microscope(s) you are using for each objective lens using the space at the bottom of the Label Your Microscope worksheet.

9. Explain to the participants that they will now get to practice preparing and viewing a slide through their microscope.

10. Direct the participants to move to their microscopes in their small groups. Indicate that they should locate the Microscope Practice Lab worksheet. Read through the directions with the participants as necessary.

11. Explain and demonstrate how to clean the eyepiece of the microscope between each participant.

12. Explain and demonstrate how to properly clean or dispose of the slides created during this lab.

13. If desired, you may demonstrate preparing a slide and viewing it through the microscope or have volunteers complete each step as you read through the directions once before allowing the small groups to begin work.

14. Allow small groups time to create slides and view them.

15. Depending on the time available, you may choose to offer other small objects for viewing, or allow participants to choose items in their environment to prepare as slides and view. Alternatively, a set of prepared slides may be offered for participants to view once they have successfully prepared their slides.

16. Direct participants to clean their microscopes and slides and complete any other housekeeping duties required at the end of the microscope lab.

17. Ask the Share Questions.

    - What observations did you make when you viewed the hair with no magnification?
    - What observations did you make when you viewed the hair under the magnifying glass?
    - What observations did you make when you viewed the hair through the microscope?
    - What differences did you notice?
    - (Optional) What other substances did you view through your microscope?

18. Ask the Reflect Question.

    - Why might scientists need to see objects more clearly than can be seen with our eyes?

19. Ask the Generalize Question.

    - What are some objects that you think might require a microscope to study?

20. Ask the Apply Questions.

    - Which jobs or types of study do you think require the use of a microscope?

How do you think the use of a microscope will help you learn about the world around you, either in an educational setting or in your daily life?

Variations

1. Participants may clean their slides and investigate other substances in their surroundings as time permits.

2. Participants may view previously prepared slides to practice focusing and manipulating the microscope.

Grab and Go 1 Introducing the Microscope

<table>
<tr><th colspan="2">Talk It Over</th></tr>
<tr><td rowspan="3">Share 1

What observations did you make when you viewed the hair with no magnification?

What observations did you make when you viewed the hair under the magnifying glass?

What observations did you make when you viewed the hair through the microscope?

What differences did you notice?</td><td>Reflect 2

Why might scientists need to see objects more clearly than can be seen with our eyes?</td></tr>
<tr><td>Generalize 3

What are some objects that you think might require a microscope to study?</td></tr>
<tr><td>Apply 4

Which jobs or types of study do you think require the use of a microscope?

How do you think the use of a microscope will help you learn about the world around you, either in an educational setting or in your daily life?</td></tr>
</table>

Term and Concept Discovery

Arm: part of the microscope that supports the nosepiece and lenses

Base: bottom of the microscope

Coverslip: small plastic or glass piece that covers the specimen on a slide

Eyepiece: the lens or combination of lenses in an optical instrument through which the eye views the image formed by the objective lens or lenses; ocular

Illuminator: the light source beneath the diaphragm of a microscope

Iris diaphragm: the lens or combination of lenses in an optical instrument through which the eye views the image formed by the objective lens or lenses;ocular

Microscope: an optical instrument having a magnifying lens or a combination of lenses for inspecting objects too small to be seen or too small to be seen distinctly and in detail by the unaided eye.

Nose: part of the microscope that holds the objective lens(es)

Objective lens: the lens(es) closest to the slide on a microscope

Optical lens: the lens in the eyepiece of a microscope

Slide: a usually rectangular plate of glass on which objects are placed for microscopic examination

Stage: the part of the microscope on which the slide rests

Stage clips: hold the slide onto the stage

Grab and Go 1 Introducing the Microscope

Education Standard(s)

Science Standards

8.L.2. Understand how biotechnology is used to affect living organisms. Summarize aspects of biotechnology including: • Specific genetic information available • Careers • Economic benefits to North Carolina • Ethical issues • Implications for agriculture

Bio.1.1.2 Compare prokaryotic and eukaryotic cells in terms of their general structures (plasma membrane and genetic material) and degree of complexity Unpacked: Bio.1.1.2 Proficiently use proper light microscopic techniques as well as determine total power magnification. The purpose is to use microscopes to observe a variety of cells with particular emphasis on the differences between prokaryotic and eukaryotic as well as plant and animal cells. While participants are not expected to understand how scanning and electron transmission microscopes work, they should recognize that they reveal greater detail about eukaryotic and prokaryotic cell differences.

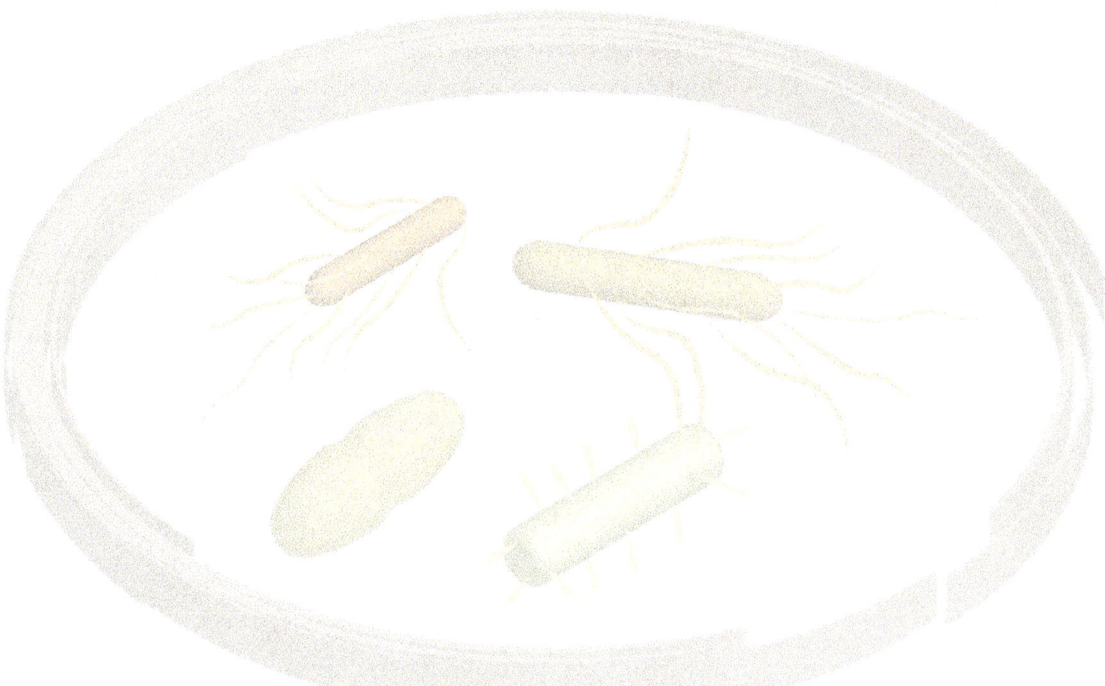

Grab and Go 1 Print Materials

Facilitator Print Materials
Label Your Microscope answer sheet

Group Handouts
Microscope Practice Lab worksheet

Youth Handouts
Label Your Microscope worksheet

*See page 2 for access to digital assets & extra enrichment.

Label Your Microscope Answer sheet

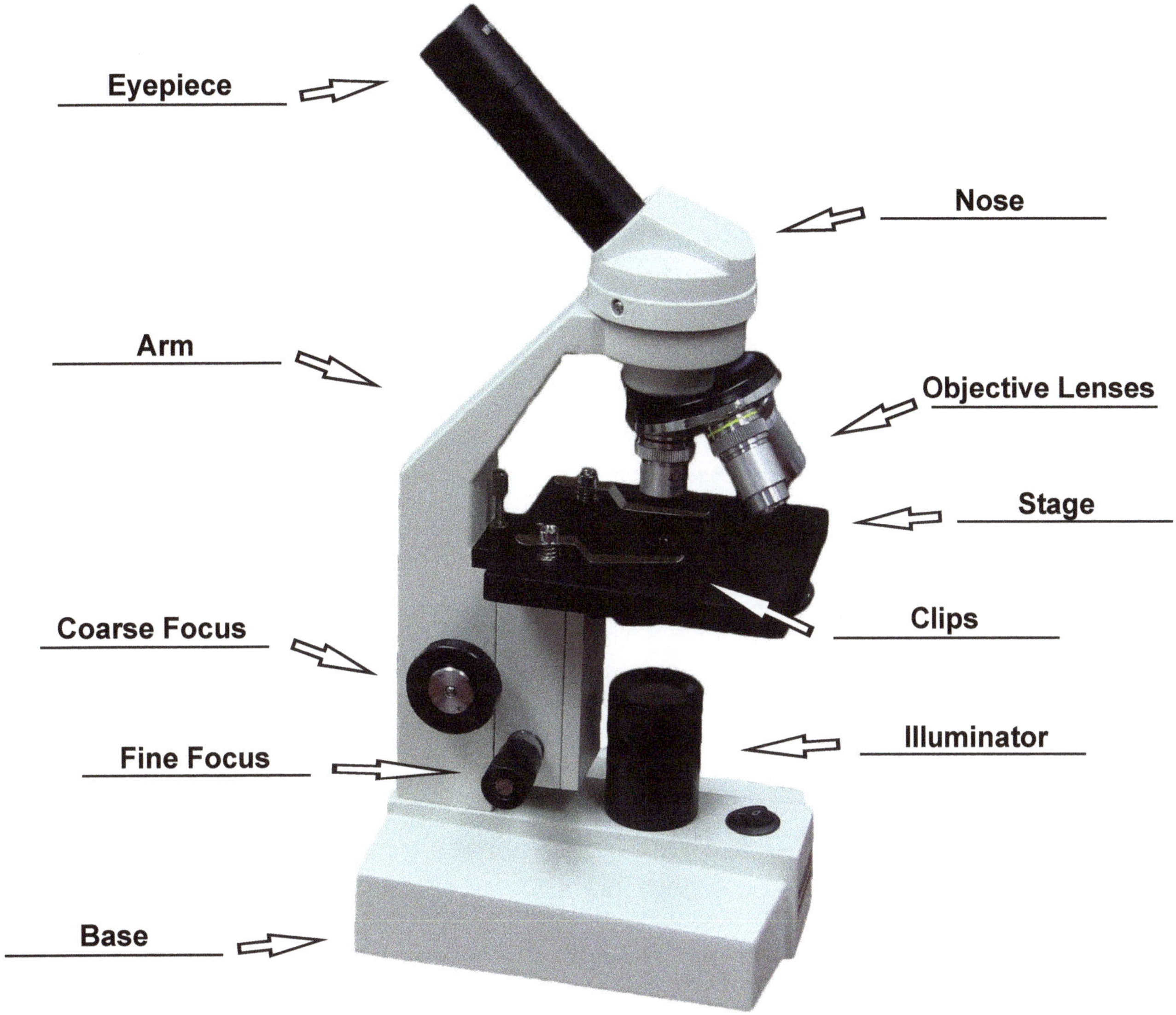

What's my magnification?
Calculate the total magnification of each lens on your microscope below:

Ocular Lens	x	Objective Lens	=	Total Magnification
______	x	______	=	______
______	x	______	=	______
______	x	______	=	______

These answers will depend on the microscopes used in the lesson.

Microscope Practice Lab

Group names: ______________________________ Date: __________

Materials you will need:

Microscope	Magnifying glass
Slides	Coverslips (or additional slides to act as coverslips)
Dropper	Small cup of water
Paper towels or napkins to use to absorb water	Alcohol swabs to clean the eyepiece between each participant
A piece of your hair!	

Observing Your Hair

1. Select a piece of hair and look at it without magnifying it in any way. Notice the color, texture, and any other details.
2. Observe the same piece of hair using a magnifying glass. Notice the same details (color, texture, etc.).
3. Put your hair in a safe place while you prepare your station to create a slide.

Making Your Slide

1. Look at your station. Identify all of the materials you will need for this lab.
2. Identify the parts of the microscope at your station. Operate the moving parts until you are familiar with how it works. Be careful!
3. Select a clean slide. Place it on the workspace in front of you.
4. Add a small amount of water (a drop or two) to your slide using the dropper.
5. Place the same piece of hair on the slide in the water.
6. Cover your slide with a coverslip or another slide.
7. Clean any extra water off of the slide with the edge of a paper towel or napkin, especially any that drips off the edge.

Viewing Your Slide

1. Rotate the coarse adjustment knob so that the stage moves down and away from the objective lenses as far as it can go.
2. Rotate the nosepiece of the microscope so that the lowest power objective lens is over the iris diaphragm (this should also be the shortest lens).
3. Place your slide on the stage and center the hair over the hole in the stage.
4. Use the stage clips to secure the slide in place. If you are using a second slide as your coverslip and the stage clips disturb your slide, you may decide not to use the clips for this lab.
5. Using the coarse adjustment knob, move the stage all the way up, as close to the objective lens as it will safely move.
6. Look through the eyepiece. Use the coarse adjustment knob and slowly move the stage down until the hair comes into focus.

Microscope Practice Lab Continued

7. If you move the stage all the way down and do not ever see the hair, move your slide on the stage and repeat steps 5 & 6 again.
8. Once the hair comes into coarse focus, use the fine adjustment knob to bring it into sharper focus.
9. Carefully rotate the nosepiece so that the next higher power objective lens is now pointed at your slide. Do not move either the focus knob or the slide during this step.
10. Move the fine adjustment knob as necessary to bring the hair into sharp focus again.
11. Notice the same details (color, texture, etc.).
12. Sanitize the eyepiece of the microscope with an alcohol swab before the next member of your group uses the microscope.
13. Carefully remove your slide from the stage and clean or dispose of it according to the instructions you are given.

Optional: Your facilitator may provide other materials for further slides or prepared slides for you to view at this time.

Label Your Microscope

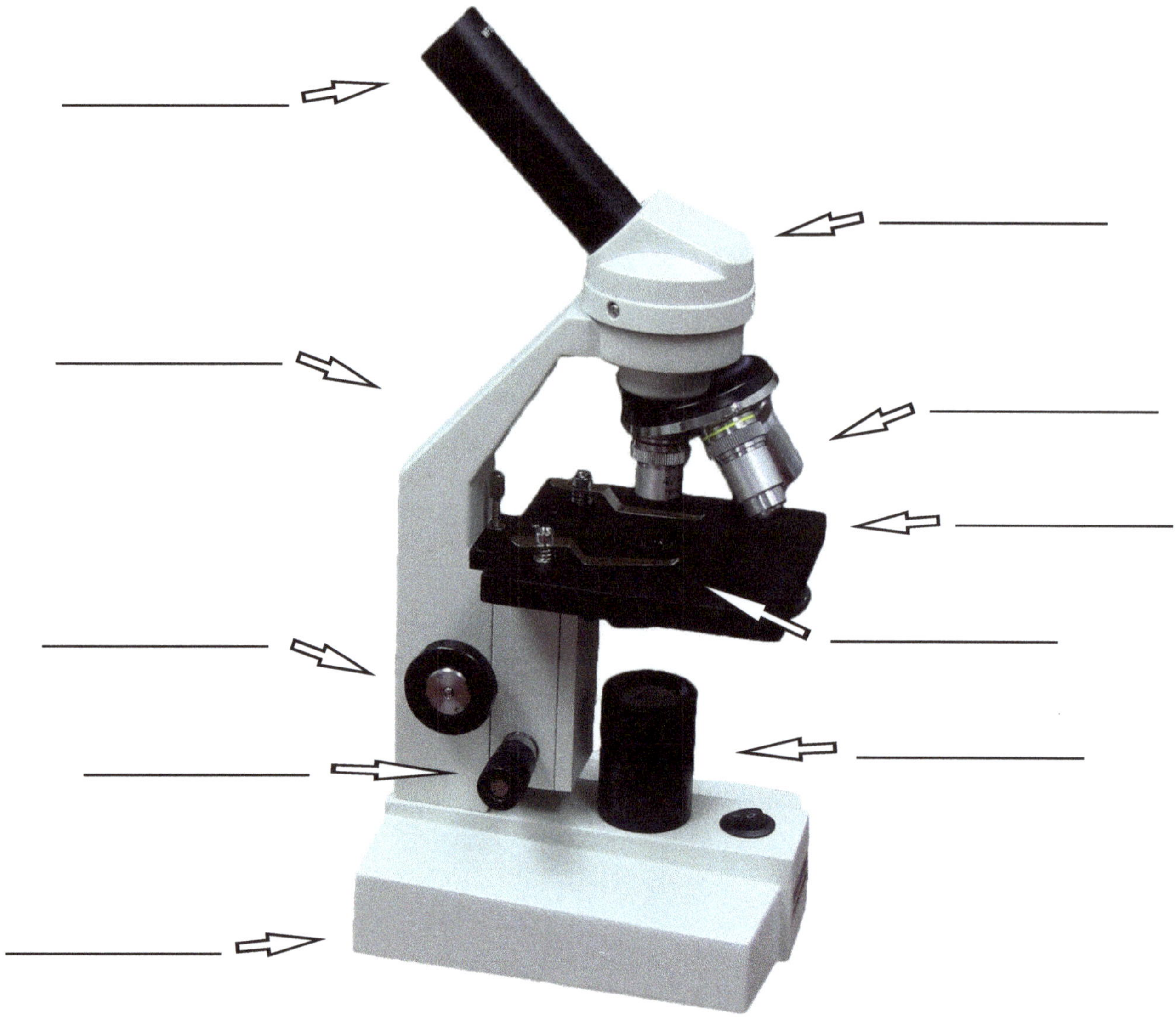

What's my magnification?
Calculate the total magnification of each lens on your microscope below:

______ Ocular Lens	X	______ Objective Lens	=	______ Total Magnification
______ Ocular Lens	X	______ Objective Lens	=	______ Total Magnification
______ Ocular Lens	X	______ Objective Lens	=	______ Total Magnification

Grab and Go 2
Outbreak Connection: Investigating Salmonella

Introduction

Salmonella is a bacteria that can cause illness in human beings. There are many different types of salmonella bacteria. Salmonella causes diarrhea and stomach problems if an individual gets infected. There are over 40,00 reported cases in the United States each year and about 400 people die each year from salmonella, so it is very important to make the public aware of this bacteria.

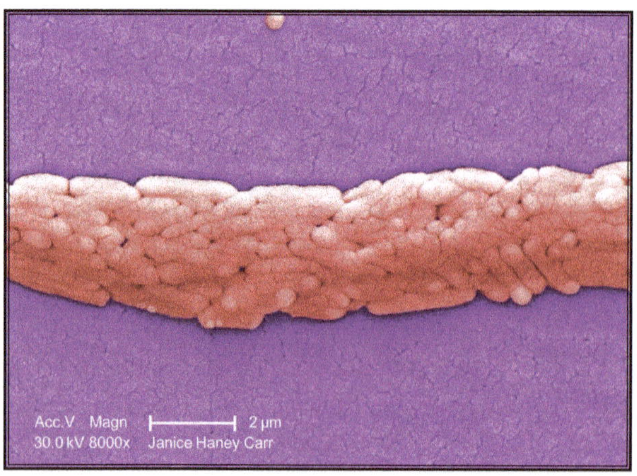

Learner Outcomes

- Develop a basic of charting salmonella outbreaks in the United States.

Skill Level: 6th-8th grade with focus on 8th grade science curriculum

Success Indicators

- Participants will use the CDC to analyze data and create a pictorial graph related to salmonella outbreaks.

Life Skills

Teamwork, Disease Prevention, Critical Thinking

Time Needed

1-2 days, approximately 60-120 minutes

Space

Classroom or area where there is ample space for all participants

Suggested Group Size

Participants will work in groups of 2-4

Materials List

- Internet access for each group
- Poster board for each group
- Markers, colored pencils, and rulers for each group

Grab and Go 2 Outbreak Connection: Investigating Salmonella

Before the Activity

- Prepare technology for participant research.
- Prepare technology to share the CDC website with the entire class at once (e.g. projector).
- Provide each group with a poster board and project supplies.

Opening Questions

- Do you know someone personally who has had salmonella?
- What do you think are some different types of places or foods that could provide an opportunity for you to contract salmonella?

Let's Do It!

1. Ask the Opening Questions.

2. Read out loud the following page from the CDC:

   http://www.cdc.gov/salmonella/general/index.html

3. Explain to the participants that their groups are all employees of the education outreach program within the CDC (Centers for Disease Control and Prevention). It is their job to analyze data, put it in youth friendly terms, and then take it to elementary school students and present to the students about salmonella.

4. Each group will analyze the following page very carefully to gather information:

   http://www.cdc.gov/salmonella/live-poultry-05-12/index.html

5. Then each group will analyze this map:

   http://www.cdc.gov/salmonella/live-poultry-05-12/map.html

6. After careful analysis and discussion within the groups, they will create presentations for elementary school students about the information that was gathered. In these presentations, the groups are required to take the outbreak map and create a pictorial graph of the data.

7. Since the students to be presented to are elementary level, the projects need to be creative and grab the attention of younger youth. Each group will create some type of graph that includes pictures. Pictures of poultry for the young students to understand may be used.

8. All presentations must include a pictorial representation graph of the current case count map, an explanation of what salmonella is, an explanation of what salmonella does to your body, and advice to the young students about how to prevent getting salmonella. BE CREATIVE! Skits, jokes, songs, etc., are all welcomed!

9. As time allows, permit each group to present either to the whole group or have small groups present to one another.

10. During the presentations, the audience will be required to give warm and cool feedback to the presenters about the presentation's suitability for younger elementary school youth. Warm feedback comments are positive reinforcement comments about how easy and relatable the presentation would be for a younger child. Cool feedback comments are constructive criticisms and ideas that could help improve the presentation and make it more relatable for a younger child. All the statements need to be directly related to facts and information in the presentation.

11. After all groups have presented at least once, ask the Generalize questions.

    - Why is it important for everyone to know about salmonella?
    - How would the education outreach department operate for the CDC?
    - What type of careers are there related to disease control?

12. If possible, have each group interact with younger youth and give their presentations.

Term and Concept Discovery

Poultry: domesticated fowl collectively, especially those valued for their meat and eggs, as chick-ens,turkeys, ducks, geese, and guinea fowl

Salmonella: any of several rod-shaped, facultatively anaerobic bacteria of the genus Salmonella, as S. typhosa,that may enter the digestive tract of humans and other mammals in contaminated food and cause abdominal pains and violent diarrhea

Acknowledgments

Information and links are all from the CDC website and permission was obtained.

Talk It Over

Share 1

Each group will do a practice presentation.

Reflect 2

During the presentations, the audience will be required to give warm and cool feedback to the presenters about the presentation's suitability for younger elementary school youth. Warm feedback comments are positive reinforcement comments about how easy and relatable the presentation would be for a younger child. Cool feedback comments are constructive criticisms and ideas that could help improve the presentation and make it more relatable for a younger child. All the statements need to be directly related to facts and information in the presentation.

Generalize 3

Why is it important for everyone to know about salmonella?

How would the education outreach department operate for the CDC?

What type of careers are there related to disease control?

Apply 4

Participants will visit younger elementary school students and do a presentation of the data analysis that they did on salmonella.

Education Standard(s)

8th grade Essential Standards:
8.L.1: Understand the hazards caused by agents of diseases that affect living organisms.

National Science Standards:
MS-LS1-5: Construct a scientific explanation based on evidence for how environmental and genetic factors influence the growth of organisms.

Tags: salmonella

Grab and Go 3 Ready Set Grow!
Examining the Growth and Spread of Microorganisms

Introduction

Microorganisms are found on the surface of almost everything. Certain microbes and other germs can be passed through human contact. This can be through skin contact or a person sneezing and the microbes landing on the next person.

In this activity participants will learn by using Glo-Germ how quickly microbes can spread and how to prevent harmful microorganisms from spreading.

Learner Outcomes

- Understand how quickly microbes can spread with human contact.

Skill Level: 6-8th grade

Success Indicators

- Participants will understand how human contact is related to the spread of microbes.
- Participants will understand the importance of hand washing and keeping their bodies healthy.
- Participants will understand the importance of food safety through proper precautions.

Life Skills

Teamwork, Critical Thinking, Communication, Contributions to Group Efforts, Disease Prevention, Problem Solving

Time Needed

One 60 minute session

Space

Classroom or room big enough for participants to move around.

Suggested Group Size

Activity will be done as a whole group. Group size of 15-25 works best.

Materials List

- Glo-Germ Gel or powder
- 1-2 black lights depending on the size of group (1 black light for 15 people works well)
- 1 bottle of hand sanitizer
- 1 bottle of hand soap
- Water source

Grab and Go 3 Ready Set Grow!
Examining the Growth and Spread of Microorganisms

Before the Activity

- Make sure there is enough Glo-Germ gel or powder for 2-3 of the participants in the group.
- If the second half of the lesson will be conducted with the whole group, make sure there is enough Glo-Germ for all participants.
- Set up the black light in an area where participants will not immediately notice it.
- Before you begin the activity with the whole group, pull two or three participants (depending on the size of the group) to the side and place some Glo-Germ gel in their hands. Make sure none of the other group members see this taking place. Have the Glo-Germ participants rub the gel into their hands. Instruct them to not tell anyone else in the group what they have just done. Have them return to the group and carry on as if nothing has happened.

Opening Questions

- How do you think certain germs or microorganisms are spread?
- What are some ways microorganisms can be slowed down from being passed between either humans or humans and other objects?

Let's Do It!

Part One

1. Ask the Opening Questions. Create a chart or list with responses from the group.

2. Have all participants stand up and tell participants they are going to "get to know" other members in the group.

3. Participants will go around the room and shake hands with at least 5 other people in the room. If this is a small group, have them shake hands with fewer people; if a larger group, have them shake hands with more people.

4. Once everyone has had a chance to go around the room and shake hands, have participants return to their seats.

5. Put on your most dramatic voice and say, "Everyone, I have some bad news…some of your classmates have a very contagious disease that can easily be spread through human contact!"

6. Give participants a few seconds to react. Ask the participants who had the Glo-Germ on their hands to stand up. Introduce them as the "contagious" members.

7. Explain to participants that anyone who came in contact with the participants with diseased hands will glow under the blacklight.

8. Have each participant place their hands under the blacklight to see if they have come into contact with the disease.

9. Once everyone has had a chance to see if they are "infected," discuss with the group how quickly the gel was able to spread.

10. Have participants who did not directly shake hands with the infected members raise their hands. See how many of them still ended up with Glo-Germ on their hands. Discuss with participants how people can still become infected even if they aren't in contact with the direct source of the disease.

11. Discuss with participants ways to prevent the spread of diseases and how this can relate to food safety. This will help lead into part two of the activity.

Part Two

This can be done with groups of 4 or with 3 individual participants.

1. Discuss with participants the importance of handwashing to help prevent the spread of microorganisms.

2. Have participants predict if they think regular water, water with hand soap, or hand sanitizer get a person's hands cleaner. This can be done as a whole group discussion or a small group discussion.

3. Have the three participants place Glo-Germ on their hands and rub it in. If you would like to do this with groups then have everyone in the group put the Glo-Germ on their hands.

4. Participants will put their hands under the blacklight to show that the Glo-Germ is present and to notice where it is located on their hands.

5. Have one person or group wash their hands for 30 seconds with regular water, have another person or group wash their hands for 30 seconds with soap and water, and have the last participant or group rub hand sanitizer on their hands for 30 seconds.

6. Once everyone has washed their hands in their assigned way, have participants place their hands back under the backlight.

7. Have participants observe which pair(s) of hands have the most Glo-Germ left on them.

8. Ask the Share question.

    - Compare your hands with other group members. Why did some people have more Glo-Germ on their hands than others?

9. Discuss with participants which form of hand-washing is better for preventing the spread of microorganisms.

10. Ask the Reflect questions.

    - What are some ways to prevent the spread of diseases?
    - What are some ways microorganisms can spread between humans?

11. Ask the first Generalize question.

    - Why do you think it is important for people to wash their hands before preparing food?

12. Discuss how it is important in the food industry for employees to wash their hands before preparing food. Have participants discuss why they think it is a law for people who prepare food to wash their hands before touching the food product.

13. Ask the Apply question.

    - What are some ways to inform people of the importance of keeping themselves away from others when they are sick?

14. Participants can wash their hands at the end if they need to remove the rest of the Glo-Germ.

Term and Concept Discovery

Food safety: a series of practices designed to minimize risks posed by microbes in the food supply. Food safety practices include proper hand washing, cooking food thoroughly, keeping perishable items chilled, cleaning food preparation surfaces and tools, and keeping different types of food separated to avoid cross-contamination.

Microbiology: the branch of biology dealing with the structure,function, uses, and modes of existence of microscopic organisms.

Microorganism: any organism too small to be viewed by the unaided eye, as bacteria, protozoa, and some fungi and algae.

Grab and Go 3 Ready Set Grow!
Examining the Growth and Spread of Microorganisms

Talk It Over

Share 1	Reflect 2
Compare your hands with other group members. Why did some people have more Glo-Germ on their hands than others?	What are some ways to prevent the spread of diseases? What are some ways microorganisms can spread between humans?
Generalize 3	**Apply 4**
Why do you think it is import for people to wash their hands before preparing food?	What are some ways to inform people of the importance of keeping themselves away from others when they are sick?

Variations

1. Any of the above activities can be done as a demonstration.
2. Make a poster explaining the importance of food safety.
3. Create a commercial showing how washing your hands with soap and warm water is more beneficial than using just water. Participants can use the Glo-Germ as part of their commercial to show the benefits of hand washing.
4. Research ways food industries take extra precaution in making sure food is prepared and packaged safely to prevent the spread of microorganisms.
5. Participants can research the spread of diseases and how outbreaks of diseases have been a huge part of earth's history.

Education Standard(s)

National Science Standards:

MS-LS1-1: Conduct an investigation to provide evidence that living things are made of cells; either one cell or many different number and types of cells

MS-LS1-5: Construct a scientific explanation based on evidence for how environmental and genetic factors influence the growth of organisms

Tags: Microbiology, disease prevention, human contact, microbe spread

Appendix B
Support Materials

Materials List
Alcohol Solution for Cleaning Goggles
Red Cabbage Juice Indicator

Material List for Middle School Group

1 group= 4 participants

Lesson 1		
Consumables	Non Consumables	Items to be printed
Antiseptic mouthwash (1 large bottle)	Microscope (1 per group if possible)	Label Your Microscope worksheet (optional - 1 per participant)
Cotton swabs (2-3 per participant)	Slides (12-14 per group)	Microscope Lab - Microbes in Your Mouth instructions (1 per group)
Toothbrushes (1 per participant)	Coverslips or additional slides	
Toothpaste (for each participant)	Computer and internet access for groups	
Floss (for each participant)	Recording device for Public Service Announcement (if available, optional)	
Small paper cups (one per participant)		
Alcohol swabs to clean the microscope eyepiece		

Lesson 2		
Consumables	Non Consumables	Items to be printed
Paper or poster (for each participant)	Access to computers or library for participant research	Wanted Poster rubric (1 per group/participant)
Glo-germ (1 container)	Blacklight	Think-Pair-Share worksheet (1 per participant)
Colored pencils, markers, rulers, etc., for all participants or groups to complete their Wanted posters		

Lesson 3 (Continued on the following page)		
Consumables	Non Consumables	Items to be printed
Plain yogurt (at least 1 cup is necessary for making yogurt to use as a starter)	Thermometer (digital works best)	Microbes in Our Food Introduction and Lab Analysis worksheets (1 per participant)
Vanilla or other flavored yogurt to taste (optional - enough for each participant) Note: Yogurt for tasting can be any flavor. The starter yogurt needs to be plain yogurt.	Permanent marker (1 per group)	Microbes in Our Food Lab Procedure (1 per group)

Material List for Middle School Group Continued

Lesson 3: Continued		
Consumables	Non Consumables	Items to be printed
Plain yogurt (at least 1 cup is necessary for making yogurt to use as a starter)	Thermometer (digital works best)	Microbes in Our Food Introduction and Lab Analysis worksheets (1 per participant)
Vanilla or other flavored yogurt to taste (optional - enough for each participant) Note: Yogurt for tasting can be any flavor. The starter yogurt needs to be plain yogurt.	Permanent marker (1 per group)	Microbes in Our Food Lab Procedure (1 per group)
Styrofoam cups with lids (1 per group)	1 Hot plate	
Plastic spoons (1 per person for yogurt bar and 2 per group when making yogurt)	1 Large (2 qt. or larger) beaker or saucepan	
Quart sized sealable bags (1 per group)	1 oven mitt for holding sauce pan	
Plastic wrap (1 roll)	1 cooler or insulated container to insulate yogurt cultures	
Newspaper (any color works)		
Napkins (enough for all participants)		
Paper cups for yogurt tasting (1 per participant)		
Toppings for yogurt tasting (these can be provided by facilitator or participants)		
Whole milk (1 gallon)		
Powdered milk (1 box)		

Lesson 4 (Continued on the following page)		
Consumables	Non Consumables	Items to be printed
Sample of water from a local water source (pond, river stream) Note: Try to get a sample that looks to be "clear with nothing in it"	Empty clear jar (mason or baby food jars work well) (1 for the whole class, 1 per group)	Microbe Data Sheet (1 per participant)

Material List for Middle School Group Continued

Lesson 4: Continued		
Consumables	Non Consumables	Items to be printed
Slide cleaning materials to use after the lab if the slides will be reused: soap and water, glass cleaner, alcohol spray, or lens cleaner **AND** paper towels, tissues, or lint-free tissues	Hand lens (1 per group)	Microscope Lab Procedures (1 per group)
	Microscope (if available, 1 per group)	
	Microscope slides (4-5 per group)	
	Coverslips for microscope slides (one slip needed per slide). (optional; additional slides may be used in place of coverslips)	
	Pipette (dropper) (1 per group)	
	Optional: Water Testing kit can be used to expand discovery of the water sample	

Lesson 5		
Consumables	Non Consumables	Items to be printed
One petri dish with agar solution (4-inch size) per participant	Subject notebook/journal and pen/pencil (1 per participant)	Microbial Math worksheet (1 per participant)
Cotton swabs (1-2 per participant)	Permanent Marker (1 per group)	Participant Data Sheets A, A-1, and A-2 (1 per participant)
Tape	Hand lenses (1 per group)	Microbes All Around Us Lab Reflection (1 per participant)
Disinfectant	Plastic pipette (1 per group)	Parent letter (1 per participant)
Safety gloves (1 pair per participant)		
Alcohol swabs to clean the microscope eyepiece		
Poster board (1 per group)		
Bleach solution		

Material List for Middle School Group Continued

Lesson 6		
Consumables	Non Consumables	Items to be printed
Coffee Filter (1 per participant)	Knife to peel any fruit as necessary	DNA Extraction Activity Participant Lab Sheet (1 per participant)
Bathroom cup (5 oz) (1 per participant)	Small test tube with lid (1 per participant)	Observation and Reflections handout (1 per participant)
Cup of warm water (tap water) (1 per participant)	Plastic Pipette (1 per participant)	
Piece of the selected item being tested (fruit or vegetable, 1 per participant, see lesson for options)	Hand lens (1 per participant)	
Plastic spoon (1 per participant)	Bottle of 91% rubbing alcohol (kept in freezer) (1 large container)	
Toothpick (1 per participant)	Colored pencils (for each group)	
Quart sized sealable plastic bag (1 per participant - consider having extra plastic bags in case any group members break their bags during the experiment)	Funnel (1 per group)	
Bottle of clear liquid dish detergent	Measuring spoon set (1 per group)	
Container of salt	Test tube brushes (for the class)	
8oz cup full of ice and water		

Grab and Go 1		
Consumables	Non Consumables	Items to be printed
Small cup of water (1 per group)	Microscope (one per group, or one for the class if more are unavailable)	Microscope Practice Lab instructions (1 per group)
Paper towels or napkins to use to absorb water	Magnifying glass (1 per group)	Label Your Microscope worksheet (1 per participant)
Alcohol swabs to clean the eyepiece between each participant	Slides (4-12 per group)	
	Coverslips (or additional slides to act as coverslips)	
	Dropper (1 per group)	

Material List for Middle School Group Continued

Grab and Go 2		
Consumables	Non Consumables	Items to be printed
Poster board (1 per group)	Internet access for each group	
Markers, colored pencils, and rulers for each group		

Grab and Go 3		
Consumables	Non Consumables	Items to be printed
Glo-Germ Gel or powder	1-2 black lights depending on size of group (1 black light for 15 people works well)	
1 bottle of hand sanitizer		
1 bottle of hand soap		
Water source		

Alcohol Solution for Cleaning Goggles

Instructions

Materials:

- 1 spray bottle to hold at least 100mL of liquid
- 70% rubbing alcohol solution
- Measuring device (graduated cylinder)
- Water
- Paper towels

Procedure:

1. Measure 40mL of rubbing alcohol and place it into the spray bottle.
2. Measure 60mL of water and place it into the spray bottle.
3. Tightly place the spray lid on the bottle.
4. Carefully invert the bottle a couple of times to mix the solution.

Application:

1. Place paper towels on a clean, dry surface.
2. Place goggles on top of the paper towels.
3. Generously spay goggles with cleaning solution.
4. Allow goggles to air dry.

Instructions for Red Cabbage Juice Indicator

You will need:

- A large red cabbage
- A large stockpot and lid
- Water
- A strainer
- An empty plastic ice cube tray
- A large plastic freezer bag

To Do:

1. Place the cabbage in the pot.
2. Fill the pot with water to cover the cabbage.
3. Place the pot on a stove burner.
4. Turn the burner on to medium-high.
5. When the water comes to a rapid boil, turn the heat down to medium.
6. Place the lid on the pot.
7. Let simmer for 30 minutes.
8. Once the liquid is good and dark, strain the cabbage leaves out. These may be composted or otherwise disposed of.
9. Pour the liquid into the ice cube tray.
10. Freeze the ice cubes overnight or until solid.
11. Pop the cubes out. Place them in the freezer bag and store them in the freezer until needed.

To use the cabbage juice indicator:

1. Thaw out a cube.
2. Add a few drops of thawed indicator to the item you are testing. You can use an eyedropper, a straw, or even a small spoon to do this.
3. If the item is acidic, the indicator will remain pink. If the item is basic, the indicator will turn light blue.

Appendix C

Teaching Tips and Techniques

Appendix C: Teaching Tips and Techniques

In this section, you will find information about some of the terms and structures used in designing the actual lessons. For information about the scientific vocabulary, please see the glossary.

The decision has been made to use the term "learners" instead of "youth" in this section, as these techniques work with adults as well.

Term	Anticipation Guides (also sometimes called Anticipation Sets)
What it is	Anticipation Guides are questions or checklists that learners complete prior to viewing or reading informational text.
How it works	Learners use their background knowledge to complete the Anticipation Guide prior to learning the information. This helps them to know what concepts they will encounter. It also helps learners to actively seek out and pay attention to the important ideas and major underlying concepts. After viewing or reading the informational text, learners go back and complete the same questions again. They are able to then compare their old knowledge with the knowledge gained as a result of the text.
When to use it	Use Anticipation Guides before and after viewing videos or reading informational text. Anticipation Guides are especially helpful for language learners who might otherwise be overwhelmed by the amount of information in a text or video.

Term	Claim-Evidence-Reasoning
What it is	Claim-Evidence-Reasoning is a lesson format that is based on scientific writing, argumentation, and thinking.
How it works	Learners encounter a research question and are asked to take a stand on a possible answer. At this point, the claim functions partly like a prediction and partly like a hypothesis, to use the language of the traditional scientific method. Learners then set about gathering evidence, which may include evidence from their personal experiences, evidence from printed or digital materials, and/or evidence from hands-on investigations. Once evidence has been gathered, students re-evaluate their initial claim. They have to decide which pieces of evidence are strongest and why. They also need to decide whether that evidence supports their initial claim, or if their initial claim needs to be revised.
When to use it	Use Claim-Evidence-Reasoning to support learners in developing scientific language and argument. Because the language mirrors the language in the Common Core State Standards, it can be used to support lessons in a variety of disciplines that require learners to identify and use evidence to support their thinking.

Term	Fishbowl Seminar
What it is	An alternative to a whole class discussion, the Fishbowl Seminar allows small group interaction and among students while involving the entire class.
How it works	The class creates two circles, an inner circle and an outer circle. The inner circle should have 6-8 seats or fewer. The facilitator asks open-ended, higher-order thinking questions. Participants on the outer circle take notes on the participants speaking in the inner circle. They are not allowed to speak or answer any questions. Participants in the inner circle discuss openly about the higher-order thinking questions that the facilitator poses. Participants in the outer circle can tap the shoulder of a participant in the inside circle when they have a strong point to make. This is how the circles change and how everyone gets incorporated into the discussion. Circle sizes can be changed to meet the needs of the group, but make sure to have more people in the outer circle than the inner circle. It is recommended to have at least 3-4 participants in the inner circle.
When to use it	The Fishbowl Seminar can be used as a class discussion tool when participant evaluation is important or to better hear each individual participant's input in a large group discussion.

Term	Four Corners
What it is	Four Corners is a formative technique for sharing out answers to a selected-response (multiple choice) question.
How it works	Learners are given a question along with several potential answers. The answers are displayed around the room, often in the corners. Learners move to the corner that matches their preferred response. Learners are allowed to share their reasons for their choice. They are also allowed to change their minds and move to a new location. A variation of Four Corners has students line up in front of their preferred choice. This creates a "human graph" which can be used to provide class data about misconceptions.
When to use it	Use Four Corners to determine what learners are thinking about key concepts and ideas in a quick, movement-based format.

Term	Formative Assessment Probes
What it is	A formative assessment probe is a question posed to gain an understanding of where learners are in their thinking. Frequently, assessment probes feature possible responses that mirror common thinking patterns among learners.
How it works	Formative Assessment Probes can be checklists, selected response questions, sorts, or other tasks that require students to articulate and apply scientific concepts. They are not used for grades; rather, they provide the base for the direction of the instruction. Learners are encouraged to explain their thinking as they complete the tasks.
When to use it	Use Formative Assessment Probes throughout a unit cycle to track learners' understanding of key concepts. Probes are especially useful at the beginning of a lesson to uncover existing misconceptions and prepare learners for the activity. The same probe can then be used at the close of the lesson to determine if learners' thinking changed or grew as a result of their work.

Term	Frayer Model
What it is	Frayer models are graphic organizer tools for developing vocabulary. It looks like this: 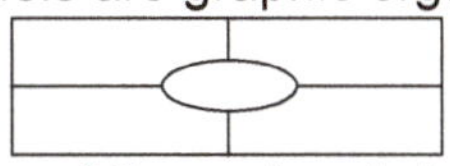 In the center of the rectangle is an oval. The new vocabulary word is written in the oval. Around the oval, the rectangular sections are labeled "Definition," "Characteristics," "Examples," and "Non-Examples." Occasionally, "Picture" or "Facts" is used in place of "Characteristics."
How it works	Learners write the new vocabulary word in the oval. Then, they complete the information in each of the rectangle sections. By having to go beyond the dictionary definition, learners develop a deeper connection to the vocabulary and a deeper understanding of the concepts.
When to use it	Frayer models can be used after new vocabulary has been introduced or as a pre-teaching tool. They are most effective once learners have had some experience with the vocabulary in context, as those experiences help learners to determine characteristics, examples, and non-examples. Frayer models can be done in class and/or as homework.

Term	Gallery Walk
What it is	A Gallery Walk is a technique for sharing ideas that have been organized in a poster, 3D model, or other highly-visual format. A variation of the Gallery Walk is the Graffiti Wall.
How it works	Learners, often in groups, create posters or 3D models in response to a question. These models are displayed, either on walls or on tables. All learners then proceed in an orderly fashion around the room, paying close attention to the work of other groups. The debrief phase of the lesson usually asks learners to summarize what they noticed during the Gallery Walk. Sometimes, learners use sticky notes to ask questions or leave feedback for the model's creator(s). In the Graffiti Wall variation, learners leave feedback or add new information directly on the original model.
When to use it	Use a Gallery Walk when learners are ready to share ideas about concepts and need a movement break.

Term	Jigsaw (sometimes called "Numbered Heads Together")
What it is	A Gallery Walk is a technique for sharing ideas and work among groups
How it works	Each group is assigned a particular component of a larger task. Group members complete the assigned component and are then reshuffled into new groups. These groups are comprised of members from each of the original groups. Together, they put the "jigsaw" together to complete the entire task.
When to use it	Use a Jigsaw to support learners as they work through large, complex tasks.

Term	"KWL"
What it is	"KWL" stands for "What I Know, What I Want to Know, and What I've Learned." KWLs are charts that learners use to track their understanding across a lesson. Variations include "TWD" (Think, Wonder, Learned) and "KWHL" (Know, Want to Know, How I Plan to Find Out, and Learned)
How it works	At the beginning of a lesson, learners complete the "K" section by listing out everything they believe to be true about a topic. Next, learners add their questions about the topic in the "W" section. The "W" section can also be used to list conflicting ideas that the class might have about a topic. Once the "W" section is complete, learners begin researching the topic through the lesson activities. Students add their new or clarified knowledge to the "L" section at the end of the lesson. During the lesson debrief at the end, students compare their ideas in "K" to their understandings in "L." They also check to see which questions were answered and which remain as topics for future investigation.
When to use it	KWL charts can be used to see what students know (or think they know) when beginning a new topic of study. They are also useful when students are going to use print or digital materials to research a topic.

Term	Stay-Stray
What it is	Stay-Stray is a technique used for sharing ideas among groups. It is similar to a Gallery Walk, but the focus is on conversation.
How it works	When it is time for groups to share their work, one group member is designated as the "speaker." The speaker stays at the group's assigned work space and is responsible for explaining the group's ideas. The other group members are designated as "strayers." Strayers are responsible for visiting the other groups and learning how they approached their work. After a few minutes, all group members return to their assigned work areas. Strayers share with their group members what they learned by visiting the other stations. The speaker shares any questions or new ideas that came up in conversation with strayers from other groups.
When to use it	Use Stay-Stray as a tool for promoting conversation among learners while sharing group work results. It is especially useful when learners have been sitting and need a movement break.

Term	Think Time
What it is	Think Time is just what it says: time given to learners to think. "Think-Pair-Share," "Think-Write-Pair-Share," "Turn and Talk," and "Wait Time" are all common variations of Think Time.
How it works	After asking a question, wait at least 5 seconds before collecting student responses. This is "Wait Time." Think Time can be extended by having learners jot ideas down and/or share their responses with a partner ("Think-Pair-Share" or "Turn and Talk") prior to sharing with the whole group. Providing Think Time is important because it allows learners time to gather their thoughts and mentally practice their responses. It also allows learners to think more deeply about questions, opening up opportunities for richer discussion. Finally, it holds more learners accountable. All learners are more likely to think about the question when they know that "the answer" won't be shared immediately. Using partner talk increases the accountability even more, as all learners then need to participate in the lesson.
When to use it	Every time a question is posed.

Term	Ticket Out the Door (also known as "Exit Ticket")
What it is	Ticket Out the Door is a quick (2-3 minute) task that learners complete at the very end of the lesson.
How it works	At the very end of a lesson, learners complete a quick task, no more than 2-3 minutes in length. Common tasks include application questions that learners answer based on the lesson, sketching activities, or reflective activities. Sometimes, Tickets Out the Door ask learners to write down areas of confusion. Tickets Out the Door provide immediate, formative feedback for teachers to help plan for the next lesson. They allow teachers to see what learners have taken away from the lesson as well as what items are still confusing and need further clarification.
When to use it	Tickets Out the Door can be used when new material is introduced or learners encounter particularly difficult material in a unit. Use them any time more information is needed about learner progress.

Appendix D

Glossary of Scientific Terms

Glossary

In this section, you will find definitions for the scientific vocabulary terms that appear throughout the unit. If you have questions about terms that refer to teaching activities, please see the section entitled, "Teaching Tips and Techniques."

Arm: Part of the microscope that supports the nosepiece and lenses.

Bacilli: Any rod-shaped or cylindrical bacterium of the genus Bacillus, comprising spore-producing bacteria.

Bacteria: A type of single-celled organism. Bacteria are classified as prokaryotic organisms, as they do not have membrane-enclosed nuclei or other similar structures (organelles). Bacteria are the most abundant life forms on our planet. Singular: bacterium

Base: Bottom of the microscope.

Biotechnology: The use of living organisms or other biological systems in the manufacture of drugs or other products or for environmental management, as in waste recycling: includes the use of bioreactors in manufacturing, microorganisms to degrade oil slicks or organic waste, genetically engineered bacteria to produce human hormones, and monoclonal antibodies to identify antigens.

Cocci: A spherical bacterium.

Coverslip: Small plastic or glass piece that covers the specimen on a slide.

DNA: Genetics. deoxyribonucleic acid: an extremely long macromolecule that is the main component of chromosomes and is the material that transfers genetic characteristics in all life forms,constructed of two nucleotide strands coiled around each other in a ladderlike arrangement with the side pieces composed of alternating phosphate and deoxyribose units and the rungs composed of the purine and pyrimidine bases adenine, guanine, cytosine, and thymine: the genetic information of DNA is encoded in the sequence of the bases and is transcribed as the strands unwind and replicate.

Epidemic: A disease affecting many persons at the same time, and spreading from person to person in a locality where the disease is not permanently prevalent.

Eukaryote: Any organism having as its fundamental structural unit a cell type that contains specialized organelles in the cytoplasm, a membrane-bound nucleus enclosing genetic material organized into chromosomes, and an elaborate system of division by mitosis or meiosis, characteristic of all life forms except bacteria, blue-green algae, and other primitive microorganisms.

Eyepiece: The lens or combination of lenses in an optical instrument through which the eye views the image formed by the objective lens or lenses; ocular.

Fermentation: A change brought about by a ferment, as yeast enzymes, which convert grape sugar into ethyl alcohol.

Food safety: A series of practices designed to minimize risks posed by microbes in the food supply. Food safety practices include proper hand washing, cooking food thoroughly, keeping perishable items chilled, cleaning food preparation surfaces and tools, and keeping different types of food separated to avoid cross-contamination.

Hygiene: A condition or practice conducive to the preservation of health, as cleanliness.

Illuminator: The light source beneath the diaphragm of a microscope.

Iris diaphragm: The lens or combination of lenses in an optical instrument through which the eye views the image formed by the objective lens or lenses; ocular.

Glossary Continued

Microbes: Any one of a number of organisms too small to be seen without a microscope. Bacteria, viruses, and many fungi are all considered microbes.

Microbiology: The branch of biology dealing with the structure,function, uses, and modes of existence of microscopic organisms.

Microorganism: Any organism too small to be viewed by the unaided eye, as bacteria, protozoa, and some fungi and algae.

Microscope: An optical instrument having a magnifying lens or a combination of lenses for inspecting objects too small to be seen or too small to be seen distinctly and in detail by the unaided eye.

Nose: Part of the microscope that holds the objective lens(es).

Objective lens: The lens(es) closest to the slide on a microscope.

Optical lens: The lens in the eyepiece of a microscope.

Organism: A form of life composed of mutually interdependent parts that maintain various vital processes.

Pandemic: A disease prevalent throughout an entire country, continent, or the whole world; epidemic over a large area.

Poultry: Domesticated fowl collectively, especially those valued for their meat and eggs, as chickens,turkeys, ducks, geese, and guinea fowl.

Probiotic: A usually dairy food or a dietary supplement containing live bacteria that replace or add to the beneficial bacteria normally present in the gastrointestinal tract.

Prokaryote: Any cellular organism that has no nuclear membrane, no organelles in the cytoplasm except ribosomes, and has its genetic material in the form of single continuous strands forming coils or loops, characteristic of all organisms in the kingdom Monera, as the bacteria and blue-green algae.

Protists: Any of various one-celled organisms, classified in the kingdom Protista, that are either free-living or aggregated into simple colonies and that have diverse reproductive and nutritional modes,including the protozoans, eukaryotic algae, and slime molds: some classification schemes also include the fungi and the more primitive bacteria and blue-green algae or may distribute the organisms between the kingdoms Plantae and Animalia according to dominant characteristics.

Salmonella: Any of several rod-shaped, facultatively anaerobic bacteria of the genus Salmonella, as S. typhosa,that may enter the digestive tract of humans and other mammals in contaminated food and cause abdominal pains and violent diarrhea.

Slide: A usually rectangular plate of glass on which objects are placed for microscopic examination

Spirilli: Any of several spirally twisted, aerobic bacteria of the genus Spirillum, certain species of which are pathogenic for humans.

Stage: the part of the microscope on which the slide rests

Stage clips: hold the slide onto the stage.

Virus: A microbe that replicates itself by taking over cells in a host organism. At this time, there is debate among scientists as to whether viruses are actually considered to be living organisms. Unlike pathogenic bacteria, viruses cannot be killed using antibiotic medicines.

Water sample: A sample of water from a specific source.

Glossary Continued

Water testing: Testing a sample of water from a specific source for water quality or other factors.

References

Lesson 1

Gomez, K. Strawberries and yogurt. [Online Image].
Retrieved March 8, 2017 from https://unsplash.com/@karlygomez?photo=lK1Q5RyD6tc licensed under CC0

Cann, A. J. Lactobacillus casei [Online image].
Retrieved March 13, 2017 from https://flic.kr/p/dHofYV licensed under CC BY-SA 2.0

D'Onofrio, A. Vibrio Cells 1 [Online image].
Retrieved March 13, 2017 from https://flic.kr/p/9djCHD licensed under CC BY 2.0

Eisenschenk, J. brushing-teeth [Online image].
Retrieved March 13, 2017 from https://flic.kr/p/7iLuUD licensed under CC BY 2.0

Hagedorn, M. Vegan Yogurt [Online image].
Retrieved March 13, 2017 from https://flic.kr/p/bnACHo licensed under CC BY-SA 2.0

Morffew, A. Roseate Spoonbill Feeding [Online image].
Retrieved March 13, 2017 from https://flic.kr/p/e1iY9w licensed under CC BY 2.0

Spragg, B. Slime mould [Online image].
Retrieved March 13, 2017 from https://flic.kr/p/r3c2XD Public Domain

Lesson 2

CDC/Palmer, E., PhD. Transmission electron micrograph of rubella virus [Online image].
Retrieved March 13, 2017 from https://en.wikipedia.org/wiki/Rubella#/media/File:Rubella_virus_TEM_B82-0203_lores.jpg Public Domain

CDC/Palmer, E., PhD. Yellow Fever Virus [Online image].
Retrieved March 13, 2017 from https://en.wikipedia.org/wiki/Yellow_fever#/media/File:YellowFeverVirus.jpg
Public Domain

derekryandublin. sick [Online image].
Retrieved March 13, 2017 from https://flic.kr/p/7yyWVM licensed under CC BY-SA 2.0

Frevert, U. Malaria [Online image].
Retrieved March 13, 2017 from
https://en.wikipedia.org/wiki/Malaria#/media/File:Malaria.jpg licensed under CC BY 2.5

Masback, K. influenza virus [Online image].
Retrieved March 13, 2017 from https://flic.kr/p/663svM licensed under CC BY-SA 2.0

NIAID. Yersinia pestis Bacteria [Online image].
Retrieved March 13, 2017 from https://flic.kr/p/9CjrCJ licensed under CC BY 2.0

Pan American Health Organization. Smallpox virus [Online image].
Retrieved March 13, 2017 from https://flic.kr/p/fvxKT6 licensed under CC BY-ND 2.0
Wikimedia Commons. Spanish Flu Ward, Walter Reed. [Online Image].

Retrieved March 8, 2017 from https://commons.wikimedia.org/wiki/File:SpanishFluWardWalterReed.jpg licensed under CC0

Lesson 3

Strong, B. 3079 *Bifidobacterium,* CDC/Bobby Strong. [Online Image]. Retrieved March 9, 2017 from https://phil.cdc.gov/phil/home.asp licensed under CC0

Lesson 4

A. Clindamycin [Online image]. Retrieved March 13, 2017 from https://flic.kr/p/bCNUKK licensed under CC BY-SA 2.0

Carr, J.H. 11792 Untreated water specimen containing unidentified organisms. [Online Image]. Retrieved March 9, 2017 from https://phil.cdc.gov/phil/home.asp licensed under CC0

Carr, J.H. 11789 Untreated water specimen containing unidentified organisms. [Online Image]. Retrieved March 9, 2017 from https://phil.cdc.gov/phil/home.asp licensed under CC0

Carr, J.H. 11785 Untreated water specimen containing unidentified organisms. [Online Image]. Retrieved March 9, 2017 from https://phil.cdc.gov/phil/home.asp licensed under CC0

Hagedorn, M. Vegan Yogurt [Online image]. Retrieved March 13, 2017 from https://flic.kr/p/bnACHo licensed under CC BY-SA 2.0

Hey Paul Studios. Guts [Online image]. Retrieved March 13, 2017 from https://flic.kr/p/afXQSL licensed under CC BY 2.0

Stephens, M. Goat and gone [Online image]. Retrieved March 13, 2017 from https://flic.kr/p/jCBVZ licensed under CC BY 2.0

Sven, S. A. Map-World-Middle-East [Online image]. Retrieved March 13, 2017 from https://commons.wikimedia.org/wiki/File%3AMap-World-Middle-East.png Public Domain

Lesson 5

Breakwell, Donald, Bryan MacDonald, Christopher Woolverton, Kyle Smith, and Richard Robison. "Colony Morphology Protocol." ASMscience | Colony Morphology Protocol. American Society for Microbiology, Sept. 2007. Web. 05 Jan. 2017. http://www.asmscience.org/content/education/protocol/protocol.3136

Dunn, Rob. "Belly Button Biodiversity." Belly Button Biodiversity. Rob Dunn Lab, 2014. Web. 09 Jan. 2017. http://robdunnlab.com/projects/belly-button-biodiversity/

Krulwich, Robert. "Bacteria Outnumber Cells in Human Body." NPR. NPR, 01 July 2006. Web. 09 Jan. 2017. http://www.npr.org/templates/story/story.php?storyId=5527426

Lesson 6

Genetic Science Learning Center. (2013, February 14) DNA Extraction. Retrieved January 09, 2017, from http://learn.genetics.utah.edu/content/labs/extraction/

Grab and Go 2

Carr, J.H. 10989 *Salmonella typhimurium*, CDC/Bette Jensen. [Online Image]. Retrieved March 9, 2017 from https://phil.cdc.gov/phil/home.asp licensed under CC0

Grab and Go 3

Gathany, J. 11162 Sneeze, CDC/Brian Judd. [Online Image]. Retrieved March 9, 2017 from https://phil.cdc.gov/phil/home.asp licensed under CC0

www.ingramcontent.com/pod-product-compliance
Lightning Source LLC
LaVergne TN
LVHW070910210926
848482LV00017B/1040
* 9 7 8 1 7 3 2 9 8 2 3 1 4 *